Blockchain Technology for Managers

Dr. Gerald R. Gray

Blockchain Technology
for Managers

 Springer

Dr. Gerald R. Gray
Knoxville, TN, USA

ISBN 978-3-030-85718-9   ISBN 978-3-030-85716-5   (eBook)
https://doi.org/10.1007/978-3-030-85716-5

This Springer imprint is published by the registered company Springer Nature Switzerland AG
The registered company address is: Gewerbestrasse 11, 6330 Cham, Switzerland

Preface

Introductions are dumb. Feel free to skip this one. However, if you feel like indulging me...

Blockchain technology has been getting a lot of attention for its potential for being a disruptive technology, and for good reason, which will be explored in this book. The purpose of this book is to be a guide for managers and other decision makers, and not a technical treatise on the various intricacies of the different types of blockchains. However managers need to know a bit about the technology to be able to make informed decisions when they are considering these types of technologies, and how they may impact their strategies or capabilities in the application portfolios that they manage. My goal here is to make you more knowledgeable or at least as knowledgeable as the salesperson that will be attempting to sell you a blockchain-based solution.

While you, dear reader, will not be asked to learn how to write a program or otherwise execute a technology implementation, you will need to know some things about the differences in the types of blockchains, the strengths and weaknesses of each, and the design trade-offs that were made in their creation. Not all blockchains were created equally. Thus, different types of blockchains may be more suited for different types of business problems.

This book will explore how we got here (the technologies that blockchain is built upon); the different types of blockchains, more aptly referred to as distributed ledger technology (DLT); and then look at some of the use cases where DLT has been, or could be, applied. With this newfound understanding of the technology you can then have informed conversations with the vendors that are proposing DLT solutions for your company.

Now, if you have been tracking DLT fairly closely and have a good grasp of the underpinning technologies, you may want to simply skip ahead to Section – Use Cases and Evaluation. Although it is my opinion that any manager assessing DLT should know enough to ask intelligent questions to vendors and have familiarity with the basics of the technology – to really get a good grasp on how products based on it may fit within your organizations portfolio – perhaps you prefer to learn by example (hopefully not bad ones!). Section 3 will explore some of the DLT use cases, where the technology may or may not fit, and you can judge for yourself, based on these examples, how other applications that are not explored within these pages may be of use to your organization.

Knoxville, TN, USA   Gerald R. Gray

Acknowledgments

I want to thank Michael McCabe for encouraging me to write this book. Michael became aware of the tutorial I created, which focused more on the high-level applications, geared towards decision makers, and broached the subject of turning the content into a book. So, here we are. It's his fault. You can blame him.

I want to thank my colleagues Hannah Davis, Micah Sweeny, Neil Hughes, and Alekhya Vaddiraj who have traveled down the road of discovery with me.

I also want to thank my wife Jamie. She has been my biggest cheerleader and supporter.

Disclaimer

This book will discuss numerous products and technologies, and may list some products as examples for different topics or implementations. These should not be construed as an endorsement of any product.

Contents

About the Author

Gerald R. Gray is a Senior Technical Executive at the Electric Power Research Institute (EPRI), where he works on special projects within the Information Communication and Cyber Security (ICCS) program, coordinating cross-program and cross-sector collaboration. In this capacity, Dr. Gray also participates in the development of industry standards as a member of International Electrotechnical Committee (IEC) TC57, and IEEE organizations. He is also a member of the GridWise® Architecture Council and a member of the Board of Directors for the Utility Communication Architecture International Users Group (UCAIUG). Dr. Gray earned a Masters of Administrative Sciences in Managing Information Systems from the University of Montana and a Doctor of Philosophy in Organization and Management with a specialization in Information Technology from Capella University.

Part I
Introduction & Fundamentals

In this first section some the drivers that answer the question, "Why blockchain?" will be explored. You should come away with an understanding why Satoshi Nakamoto might have wanted to keep their identity a secret. We'll also take a look at disruption and explore the different types and learn how to recognize it. Then we transition into learning some of the underpinning technologies that make distributed ledger technology work. This book is targeted for those managers that want to better understand technology so as to be able to make good decisions about where the technology fits and whether it matches a given problem at hand. It is not intended to be a deep dive into each technology. A balance is attempted to be struck that is "just" enough of a dive into each component that one can cover the basics, and be a snappy conversationalist at dinner when blockchain comes up, but attempting to avoid overwhelming the reader. If you're reading the book perhaps you will be making decision based on blockchain capabilities, not designing the blockchain itself.

Chapter 1
Why Blockchain?

Learning Objectives
- Why blockchain? Understanding the drivers
- Examining disruption – and learning how to recognize it

A conversation about blockchain, or distributed ledger technology (DLT) as shall be referred to for the rest of this book, DLT starts with Satoshi Nakamoto. No one knows who Satoshi is, or if it is even a single person, although some have claimed to be him or have claimed to have determined who he or she is.[1]

Why would the creator of Bitcoin want to keep their identity a secret? Is it possible that the author could foresee the potential disruption to fiat currencies that Bitcoin or other DLT tokens would represent? (Fig. 1.1)

What does disrupting the status quo, especially where the government is involved get you? One informative example is the Phil Zimmermen and the "Pretty Good Privacy",[2] case. PGP is a tool that allows users to send encrypted email without third parties being able to "snoop" on those conversations. Phil Zimmermen was investigated by the United States (U.S.) government for violating export regulations that limited cryptographic keys to a length of 40. (As a general rule of thumb, regardless of encryption type, the longer the key, the better.) The government eventually dropped its case but apropos to this conversation; it is a cautionary tale, not only for those that might disrupt the status quo, but especially for those that would disrupt the status quo where governments are concerned.

A more recent example is the case of Edward Snowden,[3] a whistleblower that caught the U.S. government unlawfully conducting a mass surveillance program on its own citizens, and for his own safety, at the time of this writing, is still living in self-imposed exile in Russia. These examples make it clear that disrupting the status

---

[1] https://medium.com/swlh/the-creator-of-bitcoin-satoshi-nakamoto-is-most-likely-this-guy-8723eddb517c

[2] https://infogalactic.com/info/Pretty_Good_Privacy

[3] Permanent Record, Metropolitan Books, ISBN-13: 978-1250237231

© Springer Nature Switzerland AG 2021
G. R. Gray, *Blockchain Technology for Managers*,
https://doi.org/10.1007/978-3-030-85716-5_1

Fig. 1.1 Who is Satoshi
Nakamoto?

Fig. 1.2 Discussing digital currency

quo, especially when disrupting the basis of commerce or governance, can come
with some severe ramifications.

Now, let us turn our attention to the blockchain story. Satoshi Nakamoto published a white paper, *Bitcoin: A Peer-to-Peer Electronic Cash System*[4] that was the
catalyst for the Bitcoin system that we know today. Let us take a look at the abstract
from that paper and break it down – there is a lot to unpack within it (Fig. 1.2).

> A purely peer-to-peer version of electronic cash would allow online payments to be sent
> directly from one party to another without going through a financial institution.

---

[4] https://bitcoin.org/bitcoin.pdf

That seems easy enough. Cash, but electronic. Just like actual cash, I can spend it without anyone having their hand out wanting a piece of the action.

Digital signatures provide part of the solution, but the main benefits are lost if a trusted third party is still required to prevent double-spending.

A digital signature? Ok – an electronic form of signature. Probably one that can be verified. Double-spending – so we are not allowed to spend the same cash on two different things. And again, being able to accomplish this without anyone having their hand out.

We propose a solution to the double-spending problem using a peer-to-peer network.

Peer-to-peer networks? Were not those "bad", like in the days of the original Napster? (Narrator: No, although Napster got a bad rap – it still exists as a music streaming service).

The network timestamps transactions by hashing them into an ongoing chain of hash-based proof-of-work, forming a record that cannot be changed without redoing the proof-of-work.

Network timestamps – something the server creates for me, but hash? Oh, where the computer takes an input essentially does some math and creates some crazy looking string of characters. But proof-of-work? I just want to spend my electronic cash – I do not want to do any work. Oh, the *system* does the work as part of the verification process – got it.

The longest chain not only serves as proof of the sequence of events witnessed, but proof that it came from the largest pool of CPU power. As long as a majority of CPU power is controlled by nodes that are not cooperating to attack the network, they'll generate the longest chain and outpace attackers.

Chain? This must be where the record of all the transactions is maintained.

The network itself requires minimal structure. Messages are broadcast on a best effort basis, and nodes can leave and rejoin the network at will, accepting the longest proof-of-work chain as proof of what happened while they were gone.

I see, so the chain that has all the records is public for those participating in the network. When I come back, I can just check the chain and get all caught up again. Records get transmitted but I can trust the rest of the network will carry on with its work while I am away.

A lot of ground is covered in that abstract. But you will see that each of these facets and more will be covered in the basics of blockchain and as we look at each of these underpinning technologies in turn. But first we will continue to look at some of the drivers and concerns that led to the creation of blockchain.

At its outset, Bitcoin was criticized for some of the various capabilities that made it attractive to its early adopters, namely, the fact that it was *pseudo-anonymous*. Nominally, Bitcoin is anonymous. The transactions are linked only to an electronic address. You do not have to share your identity. But often people or organizations share additional information that can lead to their identification, or the information is "leaked" by other means, for example, cookies or web trackers, for example, the kind used to keep track of purchases or shopping carts on a web site. Researchers found that 53 out of the 130 merchants studied leaked such information.[5] Another mechanism for being identified is that if your Bitcoins are exchanged fiat currency, to retrieve that currency you must be identified through a banking organization to collect your money.

One criticism of blockchain is that if the transactions and accounts were anonymous, then this form of currency would be attractive to bad actors. Such was the case with an online marketplace called "The Silk Road".

Silk Road was launched in 2011 by Ross Ulbricht. Ross was quoted as saying, "people should have the right to buy and sell whatever they wanted to long as they weren't hurting anyone else".[6] In addition to Bitcoin, Silk Road relied on a tool called "The Onion Router" (TOR). TOR uses a network of computers that anonymizes the route your data takes to and from the web site you may be using. Thus, the traffic cannot be traced to your location. Of course, with these kinds of capabilities it did attract the attention of people using Silk Road to buy and sell drugs. This in turn attracted the attention of federal authorities in the United States. Ulbricht was eventually arrested in 2013 while he was logged into an administrator account. He was also in the possession of 144,000 BTC (although worth much less in 2013, as of this writing, that is worth ~1.4 billion USD). Ulbricht was ultimately sentenced to two life sentences plus 40 years without the possibility for parole, for what was ultimately a nonviolent crime. This again suggests evading what the government views as its prerogatives is a path fraught with peril.

Now, did this notoriety give Bitcoin a bad name? Perhaps in the short term. But it also brought Bitcoin into the public conscious. Bitcoin seems to have survived both this and the rise, fall, and rise again of its market (in part due to currency speculators).

Why Blockchain?

Why a blockchain? In a word: shenanigans. Think back to the financial crises of 2008. Home loans were incentivized to be given to people with poor credit. Or home loans that had flexible low interest entry points that later ballooned. When the monthly house payment increased, many people found they could not afford these payments. These bad loans were bundled up into bonds, and then the bonds sold. Insurance companies underwrote the whole thing. But when people defaulted on their loans and when the bonds became worthless, and the insurance companies ran out of money to prop up the banks, the whole thing threatened to collapse. There was a lack of transparency in the process, and a lack of accountability. This was all

---

[5] https://www.technologyreview.com/s/608716/bitcoin-transactions-arent-as-anonymous-as-everyone-hoped/

[6] https://blockexplorer.com/news/silk-road-timeline-bitcoin-drugs-dark-web/

especially troubling given that these were the institutions within which there was supposed to be trust. The economy as a whole suffered, and governments around the world moved to prop up banking, insurance, and other affected industries. For example, to prop up General Motors, Ford, and Chrysler, the U.S. government acquired stock or provided loans to the tune of roughly 80 billion dollars[7] through 2008 to 2014.

For some, the policies and business relationships between these entities made for a situation that could be manipulated. What was needed was a mechanism that would ensure transparency, and a currency that could not be manipulated by governments, banks, or any other entity. There are also other practical drivers; consider the following quotes from the Satoshi white paper:

> Completely non-reversible transactions are not really possible, since financial institutions cannot avoid mediating disputes.

> A certain percentage of fraud is accepted as unavoidable. These costs and payment uncertainties can be avoided in person by using physical currency, but no mechanism exists to make payments over a communications channel without a trusted party.

Banks get drawn into this dispute resolution. But if you have a mechanism to automate an exchange, there is not anything to dispute. If the terms of a contract are met, the contract is executed. (Again, we will talk about this in more depth later.)

Is it any surprise that some felt a need for a currency not controlled by any single government or bank, with every transaction open for inspection? This is what we see with Bitcoin. Although not immune to currency speculators (like any other currency), it is immune to government influence as it is not controlled by any government. And being on a peer-to-peer network (to be explained in later chapters), very difficult to "kill". In the words of Naval Ravikant,[8] "Bitcoin is political insurance". The winds of politics may change and disrupt markets, but Bitcoin is immune from these types of shenanigans.

Disruptive Technology – How Do You Know? Why Do You Care?

This book uses the term "disruptive technology" that refers those technologies that are competence destroying. That is, they reduce or eliminate the need for some skill sets or technologies. This is opposed to those technologies that have sustaining innovation. Sustaining innovation is the situation, for example, when a new feature is added to an existing software program, or a new feature on a car or other consumer good. Disruptive technologies do away with a whole class of capabilities or

---

[7] https://www.thebalance.com/auto-industry-bailout-gm-ford-chrysler-3305670

[8] Indian-American entrepreneur and investor. He is the co-founder, chairman and former CEO of AngelList. He has invested early stage in over 200 companies. Naval Ravikant: Complete Profile and Meta List of All Things @Naval, https://unblock.net/naval-ravikant/

goods. This notion of disruptive innovation was developed by C. M. Christensen as he studied the computer hard drive industry and noted that when the architecture changed from eight-inch drives to five-inch drives, no manufacture survived that transition. When the architecture changed from five-inch drives to three-inch drives, only two manufacturers survived that transition, but only at great cost. Christensen discovered that these failures were not due to a lack of business school guidance to "listen to your customers", but in part since the legacy companies *had* listened to their customers.[9] But their customers wanted new features added to existing products; they could not envision something that completely did away with a product. The other thing that Christensen found was that this pattern repeated itself in industry after industry. The disruptive innovation curve is shown below (Fig. 1.3).

The difficulty to discern aspect of a disruptive technology is how to determine what is disruptive as opposed to what may simply be a potential failed technology. The roadside of technology advancement is littered with failed "better mousetraps". The venture capital community is another example where the expectation is that nine out of ten startups will end in failure. How then is a manager expected to identify potentially successful disruptive technologies?

One characteristic that Christensen identified is the ability for the new technology to create a new market. Often new technologies will initially underperform the market incumbents. Yet the technology has some feature that the existing technology does not have, or does not do well. Now consider DLT. As it began to be used for transactions, the criticism of the technology was the relatively slow time to clear.

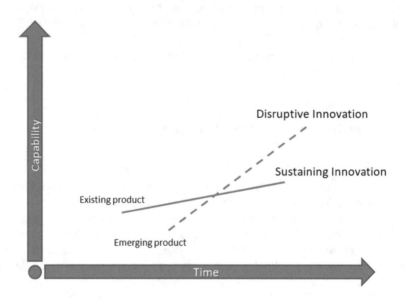

Fig. 1.3 Disruption curve. (Adapted from C.M. Christensen)

---

[9]C. M. Christensen (1997) The Innovator's Dilemma – When New Technologies Cause Great Firms to Fail, ISBN: 978-1-4221-9602. Harvard Business Review Press. Boston, Massachusetts

For example, the original Bitcoin system was designed to add a block to the chain roughly every ten minutes. For people used to the speed of the Internet, where credit card transactions clear in seconds (in fact, thousands do per second), it seemed like the system would not be able to scale. But the scalability is a valid criticism, but the early adopters likely were not adopting Bitcoin for its transaction speed, but rather the other features such as pseudo-anonymity, security, and resistance to manipulation by governments.

The other characteristic that can be seen in the disruption curve is represented by where the emerging product sales (dashed line) cross the line for the incumbent (solid line). What happens is that while originally underperforming incumbents in some measures, as the new technology is improved to address those gaps, suddenly the new technology is outperforming the old, not just in new features, but based on the old metrics as well. The new entrant eats the market from the bottom. And the new entrant is usually pretty hungry!

One of the mechanisms of creating a new customer base that Christensen also identified was how incumbents behaved relatively to the newcomers. If you are an incumbent company, perhaps grossing revenues in the billions, and some upstart shows up that is chasing after some relatively small sale, measured in thousands, those small deals do not really interest the big company. They do not have time for those. But these initially small deals mean everything to the startup. That is how they carve out a new market. This is how new entrants can often be more nimble that the old. By the time the incumbent realizes they are in danger, it is often too late. Now they are playing catch-up.

This is somewhat what we see play out in the DLT space. Bitcoin carved out a new market. Other types of DLT emerged with other characteristics, for example, Proof-of-Authority and Proof-of-Stake DLTs (the differences which will be discussed later), made different design choices, in one case, addressing the speed of transaction clearing.

The other thing that you will notice is how people and companies will jump on the bandwagon. It is interesting that DLT has made a splash across many industry verticals, such as finance, logistics, no doubt trying to leverage the technology before it eats their markets from the bottom.

Questions

1. Who is Satoshi Nakamoto?
2. What was one of the drivers for the creation of Bitcoin?
3. Why would the creator of Bitcoin want to keep their identity a secret?
4. Should you listen to your customers?
5. How can you identify disruptive technologies before they have disrupted a market?

Chapter 2
Understanding Blockchain

Learning Objectives
- Become familiar with the history of distributed ledger technology
- Become familiar with the underpinning technologies that enabled distributed ledger technology as we know it today
- Learn about the ecosystem, the applications, systems, and groups that support the technology
- Understand the basics of DLT technology

Analogies are horrible in an argument, but they are great if you are looking to illustrate a concept for the first time. So, before we get into the technical details, let us use an analogy to frame how we got here and what the environment looks like. Have you ever played the Monopoly™ board game? Chances are you have, as it has been licensed in more than 103 countries and printed in more than 37 languages.[1] With monopoly, players take turn moving around the board, buying and selling properties and collecting rents. One player acts as the banker. But if your family or friends were anything like my family and friends, you need to watch the banker like a hawk. You never knew when that person was going to start slipping $500 bills into their stack.

Now think about this situation in the corporate context and the typical corporate accounting platform. These days the transactions are automatically enabled. They have been automated and digitized. The data is centrally controlled and centrally stored. But because you still have to watch the person in control of the bank like a hawk, think of all the controls and systems put in place to put trust that the centrally controlled ledger is true and correct, facts such as:

- Two-person policies
- The person that can create an account cannot use an account

---

[1] https://infogalactic.com/info/Monopoly_(game)

© Springer Nature Switzerland AG 2021
G. R. Gray, *Blockchain Technology for Managers*,
https://doi.org/10.1007/978-3-030-85716-5_2

- The person making the change to the system must be different than the person approving the change
- Control logs
- Auditing of those controls
- Reporting
- Backup
- And all of the rules, regulations, and regulatory constructs put in place to avoid "shenanigans".

And yet, as we saw with the housing crisis and financial meltdown, that shenanigans still occurred. The advent of DLT and peer-to-peer systems has risen, in part to the failings of centrally controlled and administrated ledgers. But, before we dive into the various technologies that DLT is built upon, let us talk briefly about the DLT ecosystem.

The DLT Ecosystem: Developers, Pools, and Exchanges

Bitcoin and other DLT offerings do not stand alone, but have an ecosystem of other entities with which they interact.

Developers Each DLT has software that is used to implement its mechanics, the internal rules and code that governs such things as how the ledger is written to, how transactions are confirmed, how consensus is achieved, how data is secured, and peers communicated with. This software is often referred to as a "codebase". The developer community for any given DLT manages any changes or bug fixes that need to be implemented.

Exchanges Just like fiat currency exchanges say, where you want to exchange dollars for Euros, exchanges exist that exchange BTC to fiat currencies. Alternatively, there are exchanges that exchange different tokens, for example, BTC for ETH (Ether). Also, just like there is speculation in fiat currencies, there is speculation in cryptocurrencies as well, which can lead to fluctuations in the value that can distort a currency market.

Pool Operators A pool is a group of miners that get together to share the work (in a Proof-of-Work-based DLT) and also share the rewards. Take the Bitcoin example. In the early days of Bitcoin, you could reasonably expect to buy a computer, load the mining software, and have a decent chance that you would be rewarded for that computational work with a Bitcoin token. However, as Bitcoin grew in popularity and more computational horsepower and large data centers were dedicated to mining, the likelihood of a single operator getting the reward became much less likely. To counteract this, pools were created. This allows smaller miners to join a larger group. Now the chance of earning an award is aggregated across the group, so the chance of getting a token is higher, but of course, the rewards also must be split across the group.

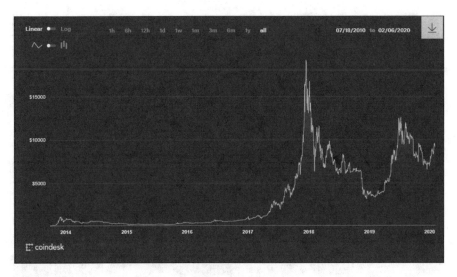

Fig. 2.1 Volatility of BTC 2014–2020. (Source: COINDESK)

Payment Processors The uptake of vendors that will take crypto tokens such as BTC or ETH has been slow. If you want to "pay the bills" in BTC, you need a payment processor that will accept cryptocurrency, and then transfer this into fiat currency. The mechanism is straightforward, but the volatility of cryptocurrencies is perhaps one reason why organizations have been hesitant to use or hold crypto tokens (Fig. 2.1).

Ethereum (ETH) has seen its own "spike" as can be seen in the figure below (Fig. 2.2).

Technology Underpinning of DLT: How We Got Here

As they say, Rome was not burnt in a day.[2] With DLT, the technology did not emerge from whole cloth in 2009; it was built upon a series of technology capabilities. While these capabilities will not be explored in depth, a manager should understand the basics of these to fully grasp any solution that they are presented with and to be able to ask pertinent questions of the presenter. Understanding the basics also provides the necessary context required for matching a product to a need and not just buying a solution, "because blockchain" (Fig. 2.3).

---

[2] https://infogalactic.com/info/Great_Fire_of_Rome

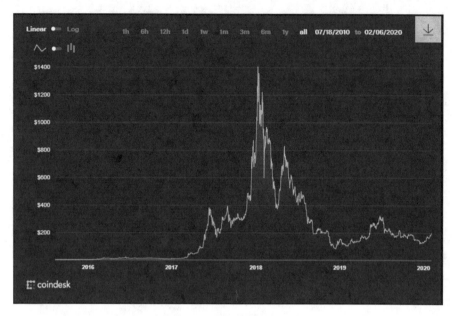

Fig. 2.2   Volatility of ETH, 2016–2020. (Source: COINDESK)

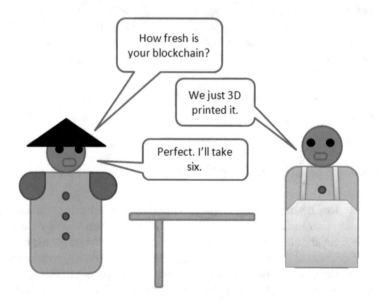

Fig. 2.3   Only the freshest blockchains

The Ledger: The Append Only Database

A standard relational database normally has the ability to both read and write data
to the database software. If you are not a computer science person or have not

worked with databases in general, you can think of a relational database as having data structured in columns and rows, just like a spreadsheet, in a database; this is referred to as a table.

Normally, unless it is protected from being overwritten, data is stored, but the data can also be edited. However, a ledger as used in DLT is append-only. This is the characteristic that is referred when DLT is described as "immutable"; that is, it cannot be changed, even if all parties wanted it to be changed (for an exception to this rule, see the section on Hard Forks).

Dr. E. F. Codd developed the relational model for database management and created tuple calculus, the inspiration behind SQL (pronounced "sequel") that facilitates querying a database for information. Hence, a relational database is tupleware.

A way to envision this is to think of Notepad, the text editor on the Windows operating system (see figure below). But imagine that each of the "records" that have been added cannot be changed. If you want to add a record, it can only be added at the end of the file. The records are sequential in time. This means, in the example in the figure below, you cannot add "record5" and put it between records three and four; it can only be added to the end (Fig. 2.4).

In a normal database, this would not be a very useful feature. Normally, based on different types of transactional systems, updates need to be performed, or corrections made. Imagine if the ledger that your bank held for your checking account contained an error, but that error could not be corrected. You probably would not be very happy. But for certain types of applications having a record that cannot be changed is an important feature. Applications such as intellectual property protection such as copyright or trademarks are examples of where you would only perhaps transfer ownership – and want an unchangeable record of that transfer. Audits are another great example. Usually, when a log or an audit of a system has been created you do not want that audit to be editable or else it loses its reliability as a record of events.

Fig. 2.4 Notepad example of an append-only file

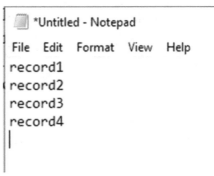

```
*Untitled - Notepad

File   Edit   Format   View   Help
record1
record2
record3
record4
```

Proof-of-Work

Proof-of-Work (PoW) was originally designed as a means to deter denial of service (DoS) attacks. A DoS attack is an attempt by an attacker to overwhelm a server with requests, essentially shutting down the server and preventing it from responding to legitimate requests. The term "Proof-of-Work" was first coined by authors Markus Jakobsson and Ari Juels.[3] The idea is that the system that makes a request has to solve a computational puzzle before the request will be serviced. But to be feasible, the puzzle must be easy to verify for the system attempting to protect itself, and at least harder for the systems making the request.

A PoW strategy was also developed in an attempt to stop spammers. The idea being that before email could be sent the sender had to satisfy a computational puzzle making it non-cost-effective to send a lot of email. Based on the proliferation of spam operations, it is arguable whether this scheme was successful in that purpose. However, for the purposes of DLT, PoW was selected as the hashing mechanism for the puzzle associated with creating a block in blockchains such as Bitcoin and Ethereum.

The development of PoW schemes was then extended by a computer scientist, Hal Finney, who developed a reusable PoW – and then to used it to make "token money". This called "HashCash" and was the first form of digital currency.

The PoW solution proposed by Satoshi Nakamoto is described thus:

> If the majority were based on one-IP-address-one-vote, it could be subverted by anyone able to allocate many IPs. Proof-of-work is essentially one-CPU-one-vote. The majority decision is represented by the longest chain, which has the greatest proof-of-work effort invested in it. If a majority of CPU power is controlled by honest nodes, the honest chain will grow the fastest and outpace any competing chains. To modify a past block, an attacker would have to redo the proof-of-work of the block and all blocks after it and then catch up with and surpass the work of the honest nodes.
>
> Bitcoin: A Peer-to-Peer Electronic Cash System, p. 3

Digital Currency

Digital currencies such as HashCash, or Bitcoin, which came after, are designed to be like gold in the sense that the value is tied to a specific, rare supply. For example, as of this writing, there are ~18 million Bitcoins in circulation, out of a total of ~21 million that will ever exist. Bitcoin uses a PoW scheme that is based on Hashcash.[4]

---

[3] "Proofs of Work and Bread Pudding Protocols". Communications and Multimedia Security. Kluwer Academic Publishers: 258–272

[4] Bitcoin: A Peer-to-Peer Electronic Cash System, Satoshi Nakamoto, p3.

The core facets of digital currencies is that they use cryptography to maintain the security of the transactions, the parties to those transactions, and also control the mechanism by which new tokens (the units of currency being exchanged) are created.

Why 21 million? A number had to be chosen. And 21 is a 'triangular' number (if you stack blocks upon each other 4, 5, 6… it creates an equilateral triangle – although it is debatable if that was the driver. It also uses a floating point number which makes for easier calculation.[5]

Another facet of these cryptocurrencies is decentralized control. Instead of being managed by a central bank working in conjunction with a government – the work of maintaining the currency is distributed among the computer systems in the networks.

Finally, for the system to work, there needs to be a mechanism to verify the history. So not just that a transaction occurred, but *when* it occurred so that it can be verified that those same coins were not also spent on something else. This is the crux of the "double-spending" problem. From the Satoshi white paper, the digital cash "chain of custody" is shown in the figure below (Fig. 2.5). This demonstrates how

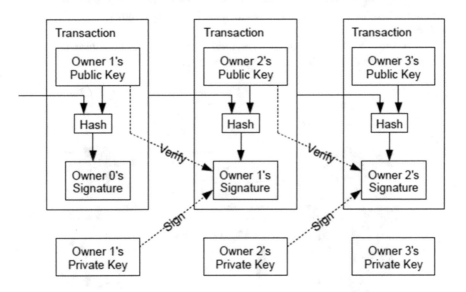

Fig. 2.5 Digital currency "chain of custody" (Bitcoin: A Peer-to-Peer Electronic Cash System, Satoshi Nakamoto)

[5]Why Did Satoshi Nakamoto Choose 21 M as Bitcoin's Maximum Supply? https://beincrypto.com/why-did-satoshi-nakamoto-choose-21m-as-bitcoins-maximum-supply/

each transaction is tied to prior transactions, creating the chain – and hence, since it is all timestamped, a mechanism to verify the history of all transactions.

Now, to sound super smart at your next manager meeting – lay this one on your peers as you are speaking with the latest DLT vendor, "I assume that the crypto mechanism your product uses is SHA-256, is that correct?" Uh oh, this manager really knows his stuff!

Hashing

Well, we better figure out what that means. SHA – is an acronym for Secure Hash Algorithm. The SHA-2 "family" includes SHA-256, SHA-384, and SHA-512. SHA-256 addresses a hash that is 2^{256} in size. Even if you are not a mathematizer, you will probably understand that this is a really big number. This family of hash algorithms was published by the National Institute of Standards (NIST) in FIPS PUB 180-2[6] and was first published in 2001. FIPS PUB 180-2 has since been superseded with new updates and capabilities, in a new family of hash algorithms in SHA-3, but SHA-256 is still used by Bitcoin. NIST latest guidance on hash functions published in 2015 can be seen table in the table below (Fig. 2.6, Table 2.1):

Ok – but how is it used? A program takes a bunch of text data, runs it through the hashing algorithm, and spits out what appears to be meaningless mix of numbers and letters.

Various forms of these hashes can be used to create digital signature, the public key-private keys that can be used as a form of identity.

Fig. 2.6 Hash – it's not just for breakfast, it's an algorithm

---

[6] https://csrc.nist.gov/csrc/media/publications/fips/180/2/archive/2002-08-01/documents/fips180-2.pdf

Table 2.1 NIST guidance on hash function (https://csrc.nist.gov/Projects/Hash-Functions/NIST-Policy-on-Hash-Functions)

SHA-1	Federal agencies should stop using this for digital signatures or timestamps. It can be used to verify old signatures and timestamps
SHA-2	This family of hash functions includes: SHA-224, SHA-256, SHA-384, SHA-512, SHA-512/224, and SHA-512/256. Application and protocol designers are encouraged to implement SHA-256 at a minimum for any applications of hash functions requiring interoperability
SHA-3	This family of hash functions includes: SHA3-224, SHA3-256, SHA3-384, SHA3-512, SHAKE128, and SHAKE256. Federal agencies *may* use the four fixed-length SHA-3 algorithms—SHA3-224, SHA3-256, SHA3-384, and SHA3-512 for all applications that employ secure hash algorithms. However, there is no requirement to move from SHA-2 to SHA-3

They can be used to create timestamps that include (hashes for timestamps), for example, Usenet Timestamping Services, and they can be used as the function to create Merkle Trees (we will explore why this is important later).

Verifiable Logs

Computers log a tremendous amount of data. All kinds of events get logged, from logins to hardware failures or other alerts. Have you ever looked at the event log on your computer? Just for fun, in Windows 10, go to the Control Panel and in the search panel type "log". You should see View Event Logs appear under the Administrative Tools like in the figure below.

Then, when the "View Event Log" link is clicked, the Event Log itself will be displayed and look something like the figure below.

In the viewer, "Application" has been selected in the panel on the left. In the center panel, various informational, warning, and error messages appear. These sorts of events are written to the log for all kinds of activities on your computer, such as application startup/shutdown, application timeouts, and so on. Any of these events can be drilled into to find additional information, in this case, for example, the application that created the event, error messages, and the timestamp of when the event was created (Fig. 2.7).

Timestamp you say? Yes. In this case, the event log provides information that can be used to troubleshoot why an error occurred and when it occurred.

Now, there might be reasons why one would want to ensure that a log had not been tampered with. One way to ensure that the log has not been tampered with is to hash the log itself. While the information we have looked in this example has a nice graphical user interface (GUI) for viewing, the contents of the log are merely text data. Thus, the text file can be hashed. Then, to ensure that when the hash for the file was created, a timestamp can be added to the hash, and this can be hashed

Fig. 2.7 Windows 10 control panel – administrative tools link

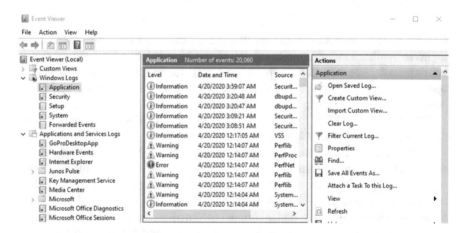

Fig. 2.8 Windows event log example

again. In this fashion, not only can the log be verified, but when it was hashed can also be verified. This is very useful if one wants to make sure that logs and time-stamps have not been tampered with in an attempt to cover one's tracks, say, in the case of a computer being hacked (Fig. 2.8).

Peer-to-Peer Networks

A distributed computing environment is fundamentally different than the prior client-server paradigm. If you are not familiar, the client server is illustrated in the figure below. You can see that the server responds to requests from one-to-many clients. The server would share its resources with the clients for services such as storage, printing, or processing data files. Generally, a client might send files to the server, but the server usually did not access files on the client (Fig. 2.9).

Peer-to-Peer networking changed that paradigm. In a peer-to-peer networking paradigm, all computers that are peers might share resources with other peers. The protocol that connects them allows for such sharing, although the administrator of any given peer could control what services would be shared with the other peers. Do you remember Napster? Napster was a music sharing service that operated from 1999 to 2001. Napster became infamous for facilitating the sharing of music files between peers. Essentially, if you had music files, you could share those, and then, if you wanted other music files, you could see what other "peers" on the network had that they had shared. Napster's argument against copyright infringement was that they did not store the music files themselves, they simply connected peers interested in sharing.

Although there were other music sharing companies that used the same peer-to-peer protocol, Napster was arguably most in the public eye for their tribulations regarding copyright infringement cases.[7] Napster faced lawsuits from music acts such as Metallica and Dr. Dre, and from the major recording labels. Eventually Napster, and other similar music sharing sites faced bankruptcy and closure under the weight of litigation. But one thing was clear, considering the millions of music files that were shared and the ease of use in which they could be shared, demonstrated the value of using peer-to-peer networking protocol (Fig. 2.10).

Fig. 2.9 Basic client-server pattern where the server "serves" requests from clients such as smart phones or other computers

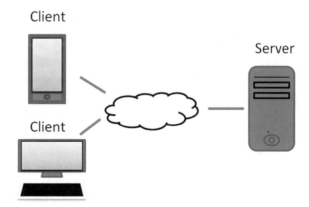

---

[7] https://www.wired.com/2010/12/mf-spotify/

Fig. 2.10 Conceptual
peer-to-peer network,
where every computer can
share information as equals

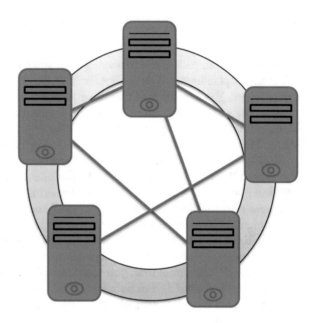

The computers connected in a blockchain peer-to-peer network work in some-what similar fashion. For example, in a PoW DLT architecture, they are all working on transactions, bundling them into blocks, and attempting to solve the "puzzle" for the next block in the chain. When one of the peers solves the puzzle, the results are sent to all the other peers for confirmation.

Byzantine Fault Tolerance

If you thought having your brother skim a few dollars from the bank when you were playing Monopoly was bad, think about the poor leader faced with the Byzantine General Problem.[89] This refers to a type of problem that can occur in distributed computing systems. This is the type of situation that you see with peer-to-peer or otherwise distributed computers. Instead of a program running on a single computer, a program may run on multiple computers that are connected via a network, and then using a protocol to send messages to each other as the program is executed.

---

[8] Lamport, L., Shostak, R., Pease, M. (1982, July). The Byzantine Generals Problem. ACM Transactions on Programming Languages and Systems. pp. 382–401. Retrieved, February 17, 2020, https://www.microsoft.com/en-us/research/publication/byzantine-generals-problem/?from=http%3A%2F%2Fresearch.microsoft.com%2Fen-us%2Fum%2Fpeople%2Flamport%2Fpubs%2Fbyz.pdf

[9] https://www.microsoft.com/en-us/research/publication/byzantine-generals-problem/?from=http%3A%2F%2Fresearch.microsoft.com%2Fen-us%2Fum%2Fpeople%2Flamport%2Fpubs%2Fbyz.pdf

Now, if the communication is disrupted, by nefarious or otherwise more innocent means, the different parts of the program need to know how to continue in the face of such a failure.

The way the Byzantine general problem is described, consider the situation where a ruler has a castle surrounded, and each of his generals is responsible for a different portion of the attack. But the attack needs to occur simultaneously or else the entire operation will fail. Normally you would think, "I'll just give a command to attack at dawn – have at you!" Unfortunately, you have gotten wind that some of your generals might not be as loyal as you would like. You cannot be sure that if you give the command that the disloyal general will follow it, and you also cannot be sure that they will faithfully pass the command onto the other generals (Fig. 2.11).

What is a troubled leader to do, when faced with a bunch of potentially disloyal generals?

Files that are being sent between systems can have a cyclic-redundancy-check (CRC). A CRC is a value that is added to blocks of data, which is based on the contents of the block. If the contents of the data block are changed, either in error or by nefarious means, then the check value will not match and an error will be produced, notifying the recipient that something has gone awry. However, to be even more certain that a message has not been tampered with a digital signature might be employed.

If you have a bunch of potentially disloyal generals (or in the DLT case, compromised nodes) and want to make sure that they are not passing false messages, the messages have an "unforgeable signature" attached to them. This is a digital signature based upon public-key cryptography and is used to make sure that the sender of a message is known to the system that is receiving the message – this is referred to as "non-repudiation". The message cannot be denied as the signature provides the authenticity. Now, keep in mind, this does not ensure confidentiality of a message – that is another issue.

Fig. 2.11 Byzantine general's problem

Public Key-Private Key Cryptography

Public key-private key is often described as a "key pair". What makes a "pair" is that the public key is generated from the private key. But what is a private key? It is simply a long, super random, number. But, the mathematics involved makes it easy to generate that public key, but in what is known as a "one-way" or "trapdoor" function, it is almost impossible to reverse the formula to get to the private key if you only have someone's public key. But they can be used for verification.

The public key is the key that is published. If it is the locking key (the key used to do the encryption), it can be used to send encrypted messages to a receiver. The private key is the secret key. It can be used to decrypt messages that were encrypted with the public key. If the private key is the locking key, it can be used to verify documents sent by the holder of the private key. Obviously, one would normally not disclose their private key, which would defeat having the private key in the first place. The public key, on the other hand, can be distributed widely.

If a message has been encrypted with the public key, only the holder of the private key can decrypt the message – thus, this ensures that a message remains confidential (Fig. 2.12).

You may hear another term used with public-key/private key encryption and that is public key infrastructure (PKI). This is a function whereby third parties certify the ownership of the key pairs; known as "certificate authorities". One common example you may be familiar with is its use with transport layer security (TLS). When you see https:// in your browser window, this indicates that the traffic to the site has been encrypted with TLS.

A digital signature, on the other hand, is signed using the sender's private key. The message that has been digitally signed can be verified by anyone that has the

Fig. 2.12 Byzantine generals passing a signed message

public key – rather than confidentiality; the purpose here is to verify that the message was sent by the holder of the private key.

Public-key cryptography is used in many types of cryptosystems and protocols. There are several ways to create the keys, which is more in-depth than what we aim for here in this book. However, the generation of all keys used in cryptography involves creating solutions to mathematical problems.

Public Keys as Identities

When you think of identity mechanisms, you may think of things such as your driver's license or perhaps your passport. These mechanisms have personally identifying information, a government issued number, your date of birth, and the period for which the identification is valid. David Chaum, concerned with how identities were being used and the potential loss of privacy, proposed a mechanism by which a public key could be used as identification, without revealing the person behind that identity.[10] In the introduction David states,

> This new approach is mutually advantageous: it actually increases organizations' benefits from automating, including improved security, while it frees individuals from the surveillance potential of data linking and other dangers of unchecked record keeping. Its more advanced techniques offer not only wider use at reduced cost, but also greater consumer convenience and protection. In the long run, it holds promise for enhancing economic freedom, the democratic process, and informational rights.

But what is a public key and how does it work? And then, how is this used as my identity?

A "key" is simply a computer-generated number. There are different mechanisms and types of keys and how they are generated, but they all have this in common. Also, a public key is derived from a private key.

A private key is what you hold. It is your "secret" key. It is still a computer-generated number. The public key is created from the private key. But you do not give out your private key, you distribute your public key when you are transacting on a system. You can used a different public key for each organization or system that you transact with. In an age when it seems, despite the supposed precautions made by vendors and banks, you still hear about data breaches where personally identifying information and credit card numbers are compromised, stolen, and then used by criminals to rack up charges or drain bank accounts. With a public key, the concern is eliminated. Even if the computer system with which you transact were to be compromised, the nefarious actor my gain your public key, but they have not gained anything that can identify you.

---

[10] Security without Identification – Card Computers to make Big Brother Obsolete, Communications of the ACM, vol. 28 no. 10, October 1985 pp. 1030–1044

Merkle Trees

"You must bring us a shrubbery… one that looks nice and isn't too expensive" – Monty Python and the Holy Grail; Knights Who Say "Ni".

In the earlier section on hashing, we discussed how DLTs such as Bitcoin use SHA to encode content using some math. But in your conversations with DLT vendors, they may also lay on you the "Merkle Tree blah blah blah", so we will touch on what this is before your eyes glaze over. First, it is not the shrubbery from the Monty Python sketch, but rather, a way to organize hashes, and hence, the information that they represent.

The "tree" can be used to verify the content and the structure of large blocks of data, for example, the kind of data that would get passed around a peer-to-peer network. The main purpose of the "tree" in this case is to make sure that the data received from other peers in the network has not changed, either intentionally or unintentionally. The "Merkle" name comes from Ralph Merkle, who patented the concept in 1979.[11]

A hash tree, such as that depicted in the figure below, is a tree of hashes (so you know it is not just a clever name – it is what it is) (Fig. 2.13).

The "leaves" are the hashes of the data blocks. The nodes show the relationship between the leaves. The leaves are related based on the child nodes. The child nodes are concatenated, and that result is also hashed.

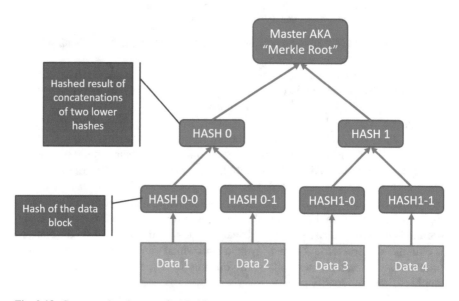

Fig. 2.13 Conceptual node map of a Merkle tree

---

[11] https://patents.google.com/patent/US4309569

Fig. 2.14 Byzantine generals discuss shruberry

Again, going back to the previous hashing discussion, a protocol like SHA-256 can be used to generate each hash based on the contents of the data block.

Thinking back the discussion on the Byzantine General Problem, if you have a message encoded with a hash, that combines data (perhaps in that scenario, movement orders and troop dispositions), that gets passed around to the other generals, if a disloyal general attempts to modify the contents, it will be flagged as it will "break"; the hash will not verify (Fig. 2.14).

Smart Contracts

What makes a contract smart? A contract requires a buyer, and seller, and consideration. Those are the basics. (I am glad I paid attention in that business law class in graduate school). A "smart" contract seeks to automate the process, such that once the terms of the agreement are met, the contract is automatically executed. One of the challenges in the past relative to smart contracts is that if I was going to use some technology with this feature, it usually meant having to use one vendors system. If one of your prospective partners did not use or have access to that system, well, too bad.

The term "smart contract" was coined by Nick Szabo.[12] Nick envisioned the use of automated contracts well ahead of their implementation. Some have even posited that Nick might be the elusive Satoshi Nakamoto.[13] Smart contracts, as envisioned

---

[12] http://www.fon.hum.uva.nl/rob/Courses/InformationInSpeech/CDROM/Literature/LOTwinterschool2006/szabo.best.vwh.net/idea.html

[13] https://101blockchains.com/who-is-nick-szabo-the-magician/

by Nick, has mechanisms to ensure that they could be verified, that they could be enforced, private, yet observable (privity, observable). Nick also wrote a paper on a concept called "Bit Gold".[14] Many of the concepts captured in that paper showed up in Bitcoin when it was developed.

These smart contracts are similar, but different than Electronic Data Interchange (EDI) in some key ways. EDI provides standards for data exchange that "provides for the exchange of business documents in a standard electronic format".[15] It was an advance in that EDI enforced a standardized format for the data being exchanged – so things like fax, mail, or purchase orders could be exchanged electronically. This mechanism is still in use by many financial institutions. However, EDI has bifurcated into multiple forms, so institutions that want to exchange EDI documents must agree on the standard and version. Data integration platforms usually support, most, if not all of the variants. Although EDI supports these types of data exchanges, it is still up to the institution to enable the appropriate security measures, and while the information related to contracting terms may be passed using EDI, the contracts cannot be executed automatically once the terms are met (Fig. 2.15).

This is an area that is ripe for some disruptive innovation. I can remember the last time I purchased a house – and the bank asked something to be… right, faxed. (What is the old quip? The 90's called and asked for their technology back). Also, sticking with our house buying analogy, if you have ever purchased a home, it seems like everyone and their best friend has their hand out; inspections, agent agreements, title insurance, more inspections, certifications, and on and on. It sure would be nice if the whole process was automated on a single platform.

Fig. 2.15 What is a smart contract?

---

[14] https://nakamotoinstitute.org/bit-gold/
[15] https://www.edibasics.com/what-is-edi/

This is the promise of using a distributed application (dApp). Again, following our real-estate analogy, if a bank was going to try to corner the "smart contract" market, they would offer the ability to automatically execute the contract and then box out their competitors. In the end, with every bank doing this, no one adopted a standard way of automating contracts. But what if a platform was offered that did not belong to any single party, but to anyone that wanted to participate? That is the promise of moving a smart contract onto a distributed application platform.

Zero-Knowledge Proof

How can you demonstrate that you know a secret, without revealing what the secret is? This is the problem addressed by the concept of the zero-knowledge proof. Also, you want the entity verifying the proof is true (that you know the secret), unable to pass on the secret information because the information is not revealed. The zero-knowledge proof is useful because it provides a mechanism to be truly anonymous – instead of simply having *pseudo*-anonymity.

Sounds complicated? Yes.

SplashData[16] posts the most common passwords that have been discovered via leaks and security breaches. Do not be these people!

This technology, like several others that we have explored in our DLT journey, traces back to the 1980s when Massachusetts Institute of Technology (MIT) researchers first proposed this concept.[17] In his blog, *Zero Knowledge Proofs: An Illustrated Primer,*[18] Matthew Green highlights one practical application of why a zero knowledge proof is useful: The classic login. You probably have dozens of accounts at various web sites. (Hopefully your password is better than "password1", yes?) Your password is stored on a server somewhere and hopefully, encrypted. But when you login, whatever you type in is recomputed and compared to the stored, encrypted password to check for a match. But the server has now learned what your password is. Which is fine… until the day when the server is hacked.

//come back to matthew green's blog//

Three facets of zero-knowledge proof:

[16] https://www.easyrobuxtoday.org/splashdata-passwords-2019-roblox/

[17] Goldwasser, S.; Micali, S.; Rackoff, C. (1989). "The knowledge complexity of interactive proof systems" (PDF). SIAM Journal on Computing. 18 (1): 186–208. https://doi.org/10.1137/0218012. ISSN 1095-7111

[18] https://blog.cryptographyengineering.com/2014/11/27/zero-knowledge-proofs-illustrated -primer/

1. Completeness – the prover will eventually convince the questioner that they are telling the truth
2. Soundness – The questioner can only be convinced if they are telling the truth
3. Zero-knowledgeness – The questioner does not learn anything else about the prover

For an explanation we will lean on Jean-Jacques Quisquater et al, who wrote a paper, "How to Explain Zero-Knowledge Protocols to Your Children[19]". In this story a person named Ali Baba has something stolen from him, and chases the thief into a cave. The cave has a fork in it, and Ali Baba does not see which fork the thief has taken. Ali Baba picks one fork but comes to the end of it, but does not find the thief. Ali Baba, of course, thinks that the thief must have taken the other fork. But when Ali Baba searches the other fork – no thief. Over the course of many days, Ali Baba is stolen from over and over again, chases the thief into the cave, but he can never catch the thief. Something is up with this cave! (Fig. 2.16)

Finally, Ali Baba starts to think there must be a better way to catch the thief than chasing them into a cave over and over again. So he hides in the cave and waits, and waits… finally a thief shows up, says the magic word, "open sesame," the secret door opens and the thief goes through, and the door closes behind him. Ali Baba has discovered the secret of the cave.

Ali Baba experiments with different magic words and discovers that when the door opens, it connects to the other cave fork. Ali Baba experimented with magic words and found that he could replace them with new magic words. (If you have not figured this out, the "magic words" in this case will be a public key per the discussion on public key cryptography earlier in this section). Ali Baba describes his discovery in a manuscript, but only provides clues to the magic, not the actual magic words.

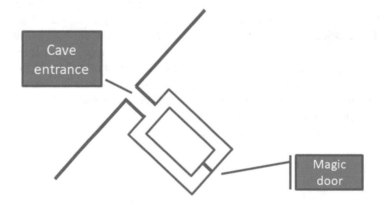

Fig. 2.16 Ali Baba and the magic cave

---

[19] http://www.cs.wisc.edu/~mkowalcz/628.pdf

However, some years later, researchers discover the cave and figure out the magic words. In the story by Jean-Jacques Quisquater, the protagonist's descendent (Mick Ali) is used in the cave experiment. News crews are invited to film both forks of the cave. Then, a reporter goes in the cave and flips a coin to decide which fork to take. Depending on whether the coin was heads or tails, the reporter yells out which fork, and of course, Mick comes out. Demonstrating that he knows the secret, without revealing what the secret is.

The paper goes on to describe how to convince others that Mick knows the secret, without revealing what it is, and how many tests can be conducted in parallel. This in essence, is the zero-knowledge path (with no math required) – a person can demonstrate that they know a secret, without revealing what the secret is. In the case of DLT, the zero-knowledge proof allows one to be truly anonymous. Or at least, extremely unlikely (how unlikely?) to be able to guess the "proof".

DLT Transaction Basics/Wallets

More vendors are beginning to accept cryptocurrency – but how does this work in practice? If I go shopping, do I hand over my bitcoins or ether? No – this is where a digital wallet comes into play, or just "wallet" in this context. Think of your own wallet. If it is like mine, it has RFID protection so that the contents cannot be scanned. An international traveler, I will often have more than one form of currency within it. My debit and credit cards now come with chips embedded within them to help secure these forms of payments. Additionally, they have two-factor authentication: Something I have (the card itself) and the PIN (something I know), when provided at a terminal, demonstrates that I have permission to use this in a transaction.

Just as a normal wallet keeps your fiat currency, a crypto wallet holds your cryptocurrency, but of course, in its electronic form, and your public keys. There are different types of wallets; some are hardware-based, but some are also software-based. For example, there are wallet apps that run on your smart phone, wallet software that you could install on your desktop computer or laptop, and web site-based wallets. And of course, wallets that work on all of these platforms and sync across devices.

There are trade-offs to using each of course. A software-based wallet, for example, one that is online, allows you to access your account from any device. While convenient, it is susceptible to the same sorts of vulnerabilities that other web-based applications have. The web sites themselves can be hacked (Fig. 2.17).

A hardware-based solution looks much like a USB stick (see figure below). They are designed so that your private keys (the ones you *do not* want to give out) stay off-line, always. That way only your public keys, the ones used in transactions, can be used.

Another view of the Ledger hardware wallet is shown below connected to a PC via the USB cable (Fig. 2.18).

Fig. 2.17 Example of a
hardware wallet. (Photo
credit: J.G. Creative
Concepts)

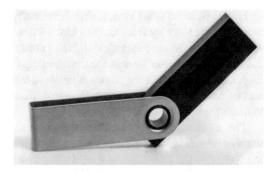

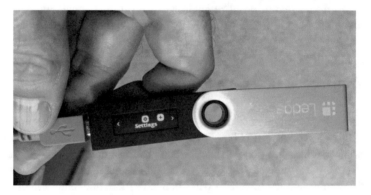

Fig. 2.18 Ledger wallet connected via USB

Let us look at some of the steps required to use a hardware wallet. These examples will use the Ledger wallet shown above.

When first acquiring the wallet, it will need to be setup with a recovery phrase and a personal identification number (PIN). The PIN works just like the PIN for your debit or credit cards. The recovery pass phrase is a bit more involved as the recovery phrase is 24 random words. The recovery phrase required for the wallet is damaged or lost. This is a downside with a hardware wallet – if you lose it and do not know your recovery phrase, it and all the cryptocurrency that you may have purchased will be gone forever. If you have a software or app-based wallet, they are only as far as your phone or computer.

Once the PIN and recovery phrase is setup, you will still need to connect to the Ledger system to install apps that will allow you to buy/sell the various cryptocurrencies. The Nano S was used in this example and it contains ~156 kb of capacity. The apps for each currency run in the 36–80 kb range so the Nano S will facilitate the buying and selling of four to six cryptocurrencies.

Ledger Live is the software that configures and works with the hardware wallet. Recall in our earlier discussion how Bitcoin was a pseudo-anonymous blockchain? Well, there is none of that here. Coinify is the platform used by Ledger to buy/sell coins and to be able to do that an account must be set up – and not only must you give your contact information, you must provide pictures of photo id, and of your-self (to prove it matches) before an account will be approved. If you like to know who you are exchanging information with – that's good. If you preferred to remain anonymous, well, this system will not be for you.

In addition to this information, Coinify also wants to know where the money came from that you are using to purchase crypto assets. An example screen is shown in the figure below (Fig. 2.19).

This seems a bit intrusive. But perhaps, they do things a bit differently over in Europe. This would not seem to be anyone's business save my own, but perhaps that is my American perspective showing through.

Once your account is set up, you will need to install an app for each cryptocur-rency that you are interested in. As noted previously, choose wisely, because this particularly hardware wallet only holds four to six apps. Once the apps are installed and accounts created for each, you are ready to buy some crypto!

The buying and selling is fairly easy. Using the menu, one simply selects the cur-rency they wish to buy (assuming they have the correct app). An example screen is shown in the figure below. In this screen the user is attempting to buy BTC using U.S. dollars. The current exchange rate is shown at the bottom of the screen. You can see at the time of the creation of this example Bitcoin was back over $26,000.

Fig. 2.19 Coinify example account setup

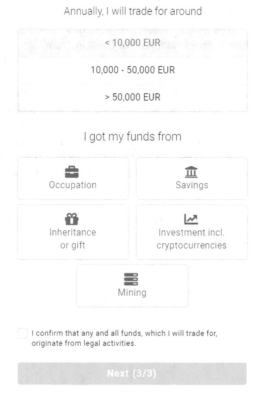

So, if you want a lesser amount, then of course you will only own a fraction of a Bitcoin (Fig. 2.20).

Coinify offers two options for paying for your crypto – you can use either a credit card (with a 4.5% surcharge) or a bank transaction (1.5% surcharge). (And you thought you were going to escape people having their handout in the crypto world).

Once you have completed your purchase, you are ready to sell it (if you want to play the money market), or use it to acquire goods and services from those that support cryptocurrency.

Ledger Data Breach

As mentioned before, software-based wallets, either smart phone app or computer-based, are subject to the same security challenges as other types of software. The systems and people are under near-constant attack attempting to find a way to get access to the keys that unlock the cryptocurrencies. However, in the following example, using a hardware-based wallet does not mean that you are immune from crypto vulnerabilities.

The customer relationship management (CRM) system of Ledger was hacked with the contact information of 272,000 customers released (see letter from Ledger CEO in the figure below).

The attacks associated with this breach came in a couple forms – phishing text messages and emails. For those not familiar with the vernacular, a phishing attack is when a hacker emulates communications from a normally trusted company. They try to get you to reveal your username/password or other credentials that would give them access to your account. The first step often involves sending a notice with links for you to click on. Following these links will result either in malware being loaded on your computer or device, or to a form where you may be encouraged to enter in your personally identifying information (PII).

In this case, it should be emphasized that unless a customer was tricked into giving away their security information related to their hardware wallet itself, the

Fig. 2.20 Buying BTC via ledger wallet

accounts and cryptocurrencies contained within it were safe. However, with the customer's PII information being leaked, now customers would be at risk of receiving targeted phishing attacks.

Some examples of the phishing emails are shown later in this chapter. There is a relative poor attempt – I wonder if someone is clever enough to configure a hardware wallet (not that it is particularly difficult) whether they would fall for the kind of attack that we see in the first email. The second email is more serious – the attackers went to some effort to make the email look much like the email that was received from the Ledger CEO.

The other risk of the PII being disclosed is that this information could be used for identity theft, where the hackers use the information to open new accounts in the customer's name. Often when other companies suffer these kinds of data breaches, customers are offered some protection services that are offered by credit reporting agencies, but Ledger did not see fit to make an offer of that nature for this incident (Fig. 2.21).

In the figure below is an example phishing email of the "not so clever" sort. The "To:" field is "You" (How did they know it was me!) (Fig. 2.22)

The "not so clever" phishing email's link goes to https://zs1a.trk.elasticemail. com/ – that does not even look "close" to ledger.com.

The next phishing email is a bit cleverer. It uses the same layout and color of the Ledger CEO's email. But again, a check of this url shows that it goes to https:// u2328921.ct.sendgrid.net/ which is also just a tad on the suspicious side.

This all goes to show, even if the wallet is secure, or even if the chain itself is secure, clicking on one of these phishing email links can break down the external security and still compromise user data. It was the breach of the Ledger CRM system that gave the hackers the phone number and email addresses – the hackers did not need to break into the Ledger wallets, if they can get the customers to hand over their coins via a malware that gets loaded onto a computer via a phishing link (Fig. 2.23).

The hacks related to Ledger have not stopped there. On January 13, 2021, the Ledger CEO sent out an additional notice that was related to the Shopify hack. Shopify[20] provides a platform for businesses or individuals to easily create an online store, complete with templates and other ease of use mechanisms to get your store started. Shopify also supports the use of transactions using cryptocurrency. The Shopify hack resulted in transaction data being stolen, which included transactional data from Ledger. The notice from the Ledger CEO is shown in the figure below (Fig. 2.24).

And the day after receiving the notice from the CEO, I received the following threatening email (figure below) "requesting" that I send someone BTC or ETH or my personal details, in this case the same sort of details that are available on

---

[20] www.shopify.com

Security Notice

Dear client,

We contacted you last July to tell you that part of our e-commerce marketing database had been leaked.

Yesterday we were informed about the dump of the content of a Ledger customer database on Raidforum. We are still investigating, but early signs tell us that this indeed could be the contents of our e-commerce database from June, 2020.
At the time of the incident, in July, we engaged an external security organisation to conduct a forensic review of the logs available. This review of the logs enabled us to confirm that approximately 1 million had been stolen as well as 9,532 more detailed personal information (postal addresses, name, surname and phone number). The database publicly released yesterday shows that a larger subset of more detailed information has been leaked, approximately 272,000 detailed information such as postal address, last name, first name and telephone number of our customers. We have previously written an FAQ for this purpose, which has since been updated.

We regret to inform you that you are part of the approximately 272 000 customers whose detailed personal information was accessed by the unauthorized third party. Specifically, your name and surname, and your postal address were exposed.

This data breach is not linked to our hardware wallets' security and your cryptocurrency funds are safe. Due to our detailed security measures, attackers cannot steal your sensitive information like your recovery phrase and private keys. You are the only one in control and able to access this information.

We deeply apologize for this security breach and are working with law enforcement to undergo an investigation

Sincerely,
Pascal Gauthier
CEO, Ledger

:II Ledger   🗑 ↻ 🐦 📘 M

Fig. 2.21 Letter from Ledger wallet CEO informing customer of the hack

 Ledger Support <ledger@sent.as>
Wed 12/23/2020 3:10 AM
To: You

Dear ,

]This is an urgent message concerning the security of your assets

We regret to inform you that our company database was recently leaked. So if you continue to use Ledger-wallet it may not be safe at this time

To ensure the safety of your funds, we recommend temporarily installing a security update in our online wallet.

Use this link to update

Fig. 2.22 The "not so clever" Ledger wallet phishing email

Dear client,

We're sorry to inform you that Ledger has fallen victim to a cyber attack and that confidential data belonging to approximately 272,000 customers has been illegally obtained by an unauthorized third party.

You're receiving this e-mail because the Ledger wallet associated with your e-mail address ( ) has been found within those affected by the breach.

To be more specific, on December 20th 2020, members of our forensics team have detected malicious software installed on one of the Ledger Live's administrative servers.

Despite our relentless efforts, as of today, it's technically impossible to make an accurate assessment of the severity of this data breach. Due to these circumstances, we must assume that your funds could be at risk of theft.

If you're receiving this e-mail, it's because you've been affected by the breach. In order to protect your assets, please download the latest version of Ledger Live and follow the instructions to set up a new PIN for your wallet.

Sincerly,
Ledger

Download latest version

Fig. 2.23 The more clever phishing email due to the Ledger hack

whitepages.com – so not a real concern. This email goes beyond phishing and makes a plausible threat, however, and this could be another way to dupe people into handing over their cryptocurrency (Fig. 2.25).

Security Notice

Dear client,

On December 23, 2020, Shopify, our e-commerce service provider, informed
Ledger of an incident involving merchant data. Rogue agent(s) of their customer
support team obtained Ledger customer transactional records in April and June
2020. This is related to the incident reported by Shopify in September 2020, which
concerns more than 200 merchants, but until December 21, 2020, Shopify had not
identified this affected Ledger as well.

We were able to examine the stolen data together with a third party forensic firm
to identify the impacted customers.

**We regret to inform you that you are part of the customers whose detailed
personal information was stolen by Shopify rogue agent(s). Specifically, your name
and surname, detail of product(s) ordered, phone number and your postal address
were exposed.**

We notified the French Data Protection Authority on December 26, 2020. We are
continuing to work with Shopify and law enforcement on the case; an investigation
is already underway, led by the FBI and the RCMP. Ledger also reported the events
to the French Public Prosecutor and filed a complaint against the rogue agent(s).

Thefts and attacks such as this cannot go uninvestigated or unprosecuted. We
continue to work with law enforcement as well as private investigators on these
cases, and we are adding more firepower by hiring additional private investigation
capacity, adding experience and approaches to finding those responsible for these
data thefts.

FINALLY, keeping you secure is our reason for existing. We will soon release a
technical solution that will remove the 24 words as the single pillar of the security
of our hardware wallets and will open the door to funds insurance.

If you would like more detail on the many steps we are taking to prevent such
incidents in the future, please read this blog post.

Sincerely,
Pascal Gauthier
Ledger CEO

Fig. 2.24 Ledger notice of Shopify hack

Software Wallet (Fig. 2.26)

Software wallets are a popular alternative to hardware-based wallets such as the
Ledger wallet due to their convenience. One popular choice is Coinbase.[21]

---

[21] Coinbase software wallet, www.coinbase.com

Cherianne Campedelli <tjmalvinagiw@outlook.com>
Thu 1/14/2021 11:16 AM
To: You

Gerald Gray
████████████████████ United States
████████

Surprised to see your personal info here?

Furthermore, you also happen to maintain a good amount of crypto. I will share all that information (and much more) with neighborhood burglars in your area.

Don't worry not yet! But, if I happen to do this, can you imagine all the possible concequences that can occur to you and your loved ones?

Scary right? But, it does not has to be that way. I will give you a way out of this.

Send me either 0.03 BTC to bc1qvfz23xkdgh9a81xp868k66chusnle2eda0nz^^4u or 1 ETH to 0x01827acFDEa97b4E2B98869f44Cc4BA^^C635fa22c [CASE-sensitive, copy & paste it, and remove ^^ from it] within the next twenty four hours, and i'll put a stop on my plan. Your private data will be removed and I will leave you alone forever.

If for any reason, you fail to meet my demand within next 24 hours, I will move forward along with my plan and whatever happens next will be on you.

I hope you will not ruin everything for yourself by making the improper choice.

Fig. 2.25 Example threatening email demanding crypto payment

Fig. 2.26 Look! I own
BTC! Will Shenanigans
follow?

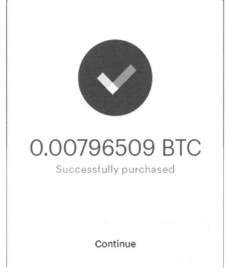

Exemplifying the ease of use, one needs only go to the web site to get started. However, just as with the Coinify option, they want to have their customers completely identify themselves before they use their app (see figure below). Coinbase does offer some convenient forms for paying for your crypto, including bank accounts, debit cards, and wire transfers. If your banking institution supports online transactions, it is fairly seamless to connect an account. And a few clicks later, voila! You have entered the world of cryptocurrency!

Coinbase allows the buying/selling of numerous forms of cryptocurrency, not just BTC. And a user can view their portfolio of their holdings from the main menu. Coinbase will also soon offers a "Coinbase card" which is a debit card tied to your Coinbase account. This sort of convenience and ease of use has made Coinbase a popular destination for those wishing to acquire cryptocurrency. It perceived value as a platform is reflected in its valuation, as of this writing, ~$68 billion, with Coinbase initially trading @ $343.58 with a planned release of 115 million shares

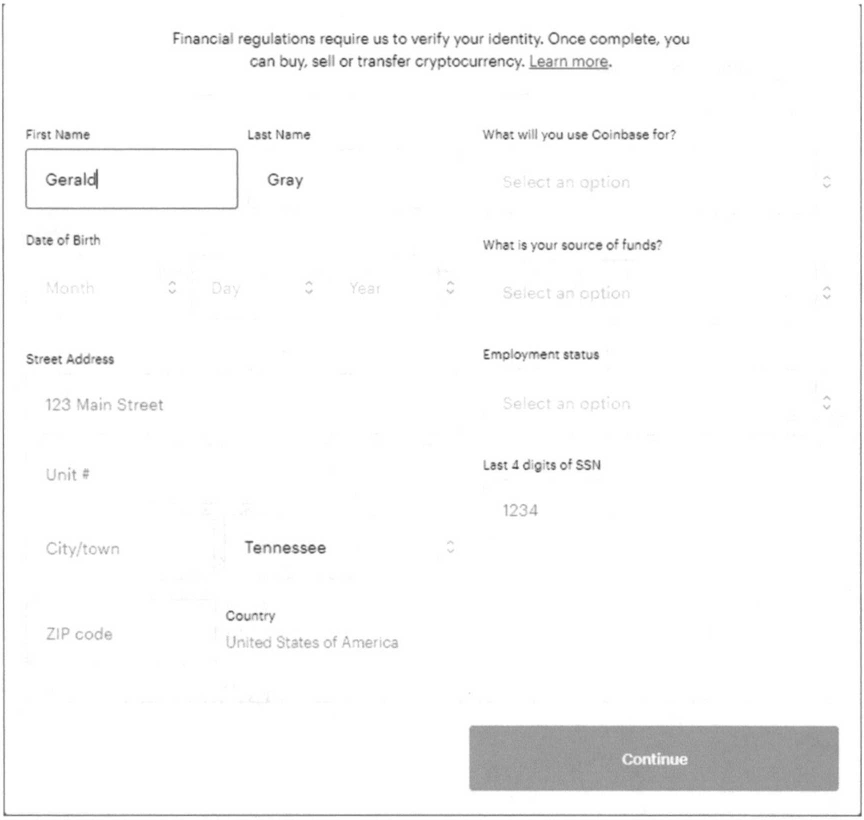

Fig. 2.27 Privacy intrusion comes standard with wallets

of common stock.[22] This definitely suggests the world of finance is changing (Fig. 2.27).

To be fair, Coinbase has had its own challenges relative to providing a target for phishing attempts. Note the image below. It *looks* like an email from Coinbase. It uses the same colors and the logo looks correct. And would not Coinbase be worried about my account (since they recently upgraded their two-factor authentication mechanism one would think so). Now the reader only has to ignore the fact that none of the links, not the sender, not the url for the "I want to disable sign-in" option, nor the "Verify My Account" button use any domain associated with Coinbase. It is just another example of how if you have questions about your account – do not attempt to address this questions by clicking on links in an email – go to the source (Fig. 2.28).

---

[22] Coinbase valuation soars to $68 billion ahead of highly anticipated crypto listing, https://www.cnbc.com/2021/03/17/coinbase-valuation-rises-to-68-billion-ahead-of-crypto-listing.html

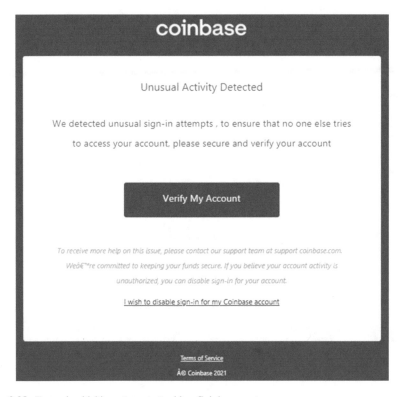

Fig. 2.28 Example phishing attempt attacking Coinbase customers

Questions

1. The four main consensus types of DLT are _____, _____, _____, _____
2. While DLT is designed to be immutable, this means that the software core will never change? True/False
3. DLT has security built-in from the "ground up", this obviously means that the ecosystem around DLT can never be hacked. True/False
4. The Bitcoin PoW process is designed to add a block of transactions approximately every ten minutes. True/False
5. Using a zero-knowledge proof allows a user to be truly _____.

Part II
Consensus, Forks, and Coins

In Part II we are going to build on the basics that were covered in Part I to better understand how those technologies are combined into the various consensus mechanisms. It is important to understand the consensus mechanisms and the resulting architecture of the various cryptocurrencies so that one has a better understanding of the design trade-offs and what implications there might be for any given use case. The most common consensus mechanisms are Proof-of-Work, Proof-of-Stake, Proof-of-Authority, and Directed Acyclic Graph, but we will take a look at a few other consensus mechanisms as well. We will also do a deeper dive into the concept of immutability and the implications of what happens when the software that runs a DLT is changed (forked in the DLT lingo).

Chapter 3
Immutability and Forks

Learning Objectives
- Blockchain hacks
- Hard forks
- Responses by the community

DLT is immutable (non-changing?) Yes. Mostly. As we discussed in the earlier section, the append-only ledger can only have records added to it. And these records cannot be edited. Does this mean that a given blockchain never changes? No.

The implementation of a given DLT may change, if the community agrees that it should change. The developers in charge of the respective code base can agree to make a change, but the change needs to be adopted by the community (miners), the peers in the distributed network that run the code of the DLT. There have been a few examples where different DLTs are needed to make a change to the software. Let us look at a couple of examples.

Ethereum Hard Fork

What happened? Everything *seemed* great. Ethereum had created the Decentralized Autonomous Organization (DAO) and raised $150 million. At the time, it was the largest crowd sale (AKA "initial coin offering") in the world.[1] The investors do not receive "shares" as one would think of in an investor-owned company, but the intent is to receive voting rights in the governance of the organization. However, the next month, the DAO was hacked and the $150 million was gone. Some "white hat" hackers were able to retrieve $100 million of the stolen funds, but then a debate

---

[1] https://medium.com/social-club/down-the-rabbit-hole-ethereum-immutability-consensus-rule-forks-e9fa8faa9e07

© Springer Nature Switzerland AG 2021

G. R. Gray, *Blockchain Technology for Managers*,
https://doi.org/10.1007/978-3-030-85716-5_3

ensued as to what to do about the hack. Some favored a "soft" fork (backwards compatible non-breaking change), while others favored a hard fork. The hard-fork folks won out and then on July 20, 2016, the following was posted to the Ethereum blog (Fig. 3.1):

> We would like to congratulate the Ethereum community on a successfully completed hard fork. Block 1920000 contained the execution of an irregular state change which transferred ~12 million ETH from the "Dark DAO" and "Whitehat DAO" contracts into the WithdrawDAO recovery contract. The fork itself took place smoothly, with roughly 85% of miners mining on the fork[2]

And with that, the "immutable" blockchain was changed. Albeit users of the original Ethereum were notified that they could continue to use the prior version, they were warned to take the necessary precautions that had necessitated the impetus for the change. The old chain is now referred to as "Ethereum Classic" (ETHC), and as of this writing, still has a market cap of over $500 million USD, (although this pales in comparison to the near $14 billion market cap for Ethereum itself). Although

Fig. 3.1 Ethereum 2016
hard fork (permission
Ethereum.org)

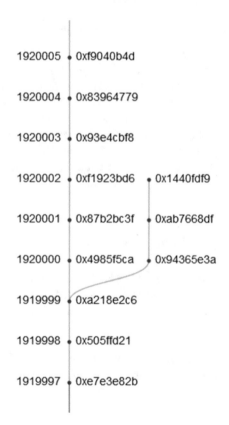

---

[2] https://blog.ethereum.org/2016/07/20/hard-fork-completed/

most have transferred to using the forked version of Ethereum, it is interesting to note that the developers created a mechanism such that for a user of Ethereum, post fork, would have the same amounts on both networks, but transactions between the two would be occasionally synced; the only exception being that the forked version of Ethereum is designed to ignore any transactions from the "Dark DAO".[3]

The DAO fork was the first unplanned fork of the Ethereum blockchain. It has since had three additional hard forks. The hard forks, names, and reasons are shown in the table below (Table 3.1).

What is the net? Even immutable things can sometimes change. The *intent* is immutability. And this was probably the biggest argument against the hard fork to correct the DAO situation. But… bugs happen. The users of DLT need to have confidence in the system with which they are transacting. This takes us back to the

Table 3.1 Ethereum hard forks since 2016

Fork	Date	Planned/ unplanned	Reason
DAO	7/20/2016	Unplanned	Correcting the hack that illegally moved $150 million from the legitimate account
EIP-150 Hard Fork	10/18/2016	Unplanned	Counter an attack that had the effect of a denial of service attack; certain transactions that were easy to request but hard to process
Spurious Dragon	11/22/2016	Unplanned	This contained for Ethereum Improvement Proposals (EIP): EIP 155 – replay attack protection EIP 160 – EXP cost increase (EXP is an "opcode" – this was done to balance the price with the complexity of the operation) EIP 161 – this allowed for the removal of numerous empty accounts – this facilitated faster sync times EIP 170 – increased the maximum code size that a contract could have; 24576 bytes
Byzantium	10/16/2017	Planned	EIP 140 – addition of a REVERT opcode EIP 196 – elliptic curve addition EIP 197 – pairing checks EIP 198 – enabling RSA signature verification EIP 211 – support for variable length return codes EIP 214 – added the STATCALL opcode which enabled non-state changing calls to other contracts EIP 649 – it delayed the difficulty change (ice age/ difficulty bomb[a]) by 1 year and reduced the block reward from 5 to 3 ether

[a]Ice age/difficulty bomb is a change to the code that would make it exponentially more difficult for miners to process blocks. This is planned to incentivize miners to move to the new version of the Ethereum blockchain. By increasing the difficulty it would make processes slow to a halt, hence, "ice age"

---

[3] https://medium.com/@timonrapp/how-to-deal-with-the-ethereum-replay-attack-3fd44074a6d8#.ocsfgea7l

community of developers that support the code base. They need to have an open and transparent process for how changes are considered, communicated, and implemented. This seems to be the case with the Ethereum community. The community posts updates at blog.ethereum.org and also has a bounty program for developers to document vulnerabilities.[4] So is DLT immutable? Sort of.

//address Ethereum classic blockchain hack//

Bitcoin Hacks and Forks

Bitcoin has seen some forks as well, usually as a result of a community desiring to see some different features, for example, changing how many transactions can be stored in a block or how many transactions per second can be supported. One of the most successful forks of Bitcoin was the creation of Bitcoin Cash in 2017. Developers were motivated to make a change because the transaction fees had gone from just pennies, to a few dollars. This would be accomplished by changing the size of a block from 1 to 8 MB, increasing the number of transactions that a block could hold.[5] Original Bitcoin processes seven transactions per second – this is what has led to scalability criticisms. Bitcoin cash can process 61 transactions per second. It is still not "Visa scale" (~1700 transactions per second) by a long shot, but it is an improvement.

For an extensive list of Bitcoin forks and their fork dates, see https://www. forks.net/list/Bitcoin/

The original Bitcoin, like other software, has seen maintenance and other features added over time. These are referred to as "soft forks" because the software update is backwards compatible with prior versions. Whereas with a hard fork there is a fundamental change to the software code that is incompatible – and anyone running the old version of the software will not be compatible or be able to process blocks generated by the new code.

Visa averages about 1700 transactions per second based on their claims of averaging 150 million transactions per day.[6] They claim to be able to handle 24,000 TPS. In 2011, according to a blog post, they were hit ~11,000 TPS during the busiest minute on December 23 of that year.[7]

---

[4] https://bounty.ethereum.org/

[5] https://www.bitdegree.org/crypto/tutorials/bitcoin-fork

[6] https://usa.visa.com/run-your-business/small-business-tools/retail.html

[7] https://www.visa.com/blogarchives/us/2011/01/12/visa-transactions-hit-peak-on-dec-23/index.html

Where bitcoins were moved – and 30% are still there.
https://medium.com/meetbitfury/crystal-blockchain-analytics-investigation-of-the-zaif-exchange-hack-a3b4d1faed8f

Bitcoin "Twitter" Hack

"I'm giving back to my fans. All Bitcoin sent to my address below will be sent back doubled" … and thus began a hack that resulted in more than $100,000 in BTC being stolen via compromised twitter accounts of celebrities and companies.[8] The Kanye West example is shown here, but hacked account included Bill Gates (retired CEO of Microsoft Corporation), Elon Musk (Tesla), and Jeff Bezos (Amazon). Basically, any person or company that one might think suddenly had $10,000,000 burning a hole in their pocket that they needed to get rid of – or maybe their conscience. But of course, this hack falls into the "if it's too good to be true it probably is" category. This goes to show that you might be smart enough to invest in Bitcoin but not smart enough to avoid an obvious scam such as this one (Fig. 3.2).

Twitter temporarily froze verified accounts to stem the tide of hacks. Twitter indicated that the hackers targeted some employees that had access to internal systems and tools.[9] As a precaution Twitter also locked the accounts of anyone that had

Fig. 3.2 Example tweet from the Twitter/Bitcoin hack

---

[8] https://www.nbcnews.com/tech/security/suspected-bitcoin-scammers-take-over-twitter-acc https://www.floridatoday.com/story/tech/science/space/2020/07/15/accounts-like-elon-musk-and--bill-gates-hacked-bitcoin-twitter-breach/5446449002/ounts-bill-gates-elon-n1233948
[9] https://twitter.com/TwitterSupport/status/1283591846464233474

changed their password in the prior 30 days in an attempt to catch any further com-promised accounts.

This was another example of not Bitcoin itself being hack, but a system that leveraged Bitcoin as a medium of transfer to garner the proceeds of the hack.

Not a Hack – Just the Government Doing Government Things

A London-based DLT analysis firm noticed that the fourth largest Bitcoin wallet, which was associated with Silk Road, had suddenly, after years of no transactions, moved an enormous amount of Bitcoin.[10] The move was of 69,369 Bitcoin, and at the time was worth approximately $1 Billion. With the recent surge in Bitcoin pric-ing, as of this writing it is worth ~$2.2 Billion.

In the twitter verse, there was a bit of a conversation going on as to whether this had represented some new large hack. Since Ross Ulbricht, the person behind Silk Road was still in prison, he probably could not have made the move.

No, it turned out that after all these years (Ulbricht sentenced in 2013), the FBI had finally confiscated the coins. Now suddenly an agency of the U.S. government became one of the single largest holders of Bitcoin. Not a hack. Nothing to see here folks. Just the government doing government things.

IOTA Hack

IOTA (see description of the Directed Acyclic Graph in the architecture section) had changed their architecture such that a node called "Coordinator" would have the final say on validation. There are trade-offs with any such single point of failure. One the one hand, it gives administrators a single point of control making it easier to manage. On the other hand, it invites disaster. With this hack, there was a little bit of both.

In this case, there was a third-party mobile and desktop wallet app called "Trinity". In February, 2020, hackers were able to take advantage of vulnerability in this app and stole over a million US dollars, from 50 accounts.[11] In response to this hack, IOTA administrators decided to shut down the entire network. This was a heretofore unprecedented action in the DLT community.

---

[10] https://www.cnbc.com/2020/11/04/1-billion-of-bitcoin-linked-to-silk-road-is-on-the-move-elliptic.html

[11] https://www.zdnet.com/article/iota-cryptocurrency-shuts-down-entire-network-after-wallet-hack/

The developer community began making changes and migrating to a patch system from February 29 through March 7, with the system migrated and restored on March 10.[12]

When There Is Risk – Insure

Into the breach of DLT hacks one might wonder, above and beyond normal security precautions, what else could be done to manage risk? Tokens, since they are not deposits like one has at the local bank, are not covered by the normal FDIC depositors insurance. To this end, some insurance companies have stepped forward with some offerings to protect one's token holdings. Coinover,[13] the Great American Insurance Company,[14] and Lloyds[15] are some examples of companies that now offer insurance products for cryptocurrency.

Like any insurance, one must conduct due diligence to understand the terms of such products and determine whether the cost is worth the risk. Although these products are for cryptocurrency, it suggests that the same type of care must be taking when considering the service-level agreements (SLA) of any dApp that runs upon a given DLT architecture.

The main takeaway from the hard fork conversation is that the code base can change, and hence, some of the transaction records that were supposed to be permanent. The changes range from benign (increasing performance), to more serious (rolling back transactions), but the rollback only occurred in the instance when it was determined that a bad actor had been able to take advantage of a flaw in the system. To date, changes have not been made for superfluous reasons. Could an outside entity coerce or influence the developer community responsible for the code base for a given DLT? Perhaps – but the "hue and cry" from the community at large would most likely result in the associated DLT, and its cryptocurrency, being dropped – immutability is a key feature of the technology. Users would most likely transfer their assets to other DLTs – by eliminating the key facet, immutability, ironically, the DLT would lose its "trust" component.

Is there risk? Of course, the ecosystem around DLT is vulnerable to the same kinds of attacks as any other software. Holders of tokens may be the victims of phishing attacks, or having their communications intercepted if the transport has not

---

[12] https://blog.iota.org/protecting-user-tokens-and-rebooting-the-coordinator-95ff96625186

[13] https://www.coincover.com/

[14] https://www.profunderwriters.com/insurance-for-cryptocurrency/

[15] https://www.insurancejournal.com/news/international/2020/03/02/559855.htm

been secured. Keyloggers that capture everything typed upon a keyboard can be installed either directly by third parties that gain unauthorized access, or unintentionally through malware. These types of attacks could passwords at risk if not the private keys that controlled access to a user's tokens. Thus, even though security is one of the core characteristics of DLT – users still need to take the same security precautions that they would take with any software.

Questions

1. What is Ethereum Classic?
2. How is a soft fork different than a hard fork?
3. If Kanye West tweeted that he would double your Bitcoin if you sent him some you would _____?
4. Was the Mt. Gox hack of Bitcoin a hack of the core Bitcoin blockchain?
5. One of the largest moves of Bitcoin was made by _____?

Chapter 4
DLT Types and Design Trade-Offs

Learning Objectives
- Understand the four main types of DLT
- Understand the design trade-offs that have been made by each
- Understand the concern over PoW energy consumption
- Understand a "stable" coin's value proposition
- Understand the landscape and potential for classic PoW mining

Understanding the Proof-of-Work Process

The first block in the Bitcoin blockchain was by Satoshi Nakamoto. This was block zero and came to be known as the "genesis" block. This block began the Bitcoin process. (See figure below for a high-level view of the Bitcoin process.) Thus, this naming convention has entered the lexicon of DLT terms, now referring to the initial block of other blockchains.

> Apparently, some fans of the Bitcoin Genesis block hold it in some sort of "cult-like" reverence,[1] donating Bitcoin to the Genesis block. This is seen as sort of a sacrifice because once something is moved into the Genesis block, it cannot be moved out again.

Bitcoin was designed to add a new block to the chain once about every ten minutes. Recall that in the ecosystem there are a group of individuals, pools, or large miners that use their computers to do the "mining" (the math that calculations hash of the new block). As part of the mining process, miners assemble the 10 minutes of

---

[1] https://www.investopedia.com/terms/g/genesis-block.asp

© Springer Nature Switzerland AG 2021
G. R. Gray, *Blockchain Technology for Managers*,
https://doi.org/10.1007/978-3-030-85716-5_4

transactions into a block to be added to the chain. Whichever miner solves the puzzle is rewarded with the Proof-of-Work. The Proof-of-Work is then sent to the other miners for verification. (Recall that the puzzle is designed to be hard to solve, but easy to verify.) The other miners verify the Proof-of-Work and then the whole process starts again (Fig. 4.1).

> In addition to the transactions that make up the block, the block also contains a hash of the prior block. This backwards referral is what makes the blocks into a chain.

You may be wondering how many transactions are stored in a block. The amount is limited by the block size, which is set at 1 MB. You can watch the blocks being added to the chain at https://www.blockchain.com/explorer. Here you can see the hash, for example, at the time of this writing; the latest block had a hard of: 00000000000000000000008aff7c02ac018f389a47f8f13f9dc48acf6c61c4ea5e

And the block contained 2855 transactions. You can also see what the transaction volume, block reward, and fee reward were for the block. In this case, 7046.19125831 BTC, 12.50000000 BTC, and 0.14898281 BTC.

In addition to adjusting the difficulty of solving the puzzle to ~10 minutes, the overall difficulty has steadily increased over time. The overall hashing difficulty has been increased as the capability of devices designed to compute hashes has also increased. This increase can be seen in the figure below. The difficulty is measured in terahashes. Terahashes is a measurement of how many hashes can be processed per second (Fig. 4.2).

Bitcoin Halving

In May 2020, the most recent Bitcoin "halving" occurred. This is a design of the protocol where every four years (or 210,000 blocks), the reward for mining Bitcoin is cut in half. With the reward cut in half, fewer new Bitcoins will enter the market. Instead of miners getting 12.5 BTC per block, they only get 6.25 per block. This is the third time that the reward has been reduced by this mechanism and the next will occur in 2024. One must also keep in mind that, by design, that the number of Bitcoin that will ever be created is capped at ~21 million. This design was created so that Bitcoin, all things being equal, would increase in value over time whereas fiat currencies decrease in value over time. "The Bitcoin protocol is not inflationary or deflationary in the long run. It is instead programmed to be *disinflationary*, culminating in a constant monetary base without change to the supply".[2]

---

[2] https://medium.com/the-bitcoin-times/stop-calling-bitcoin-deflationary-84462cb90345

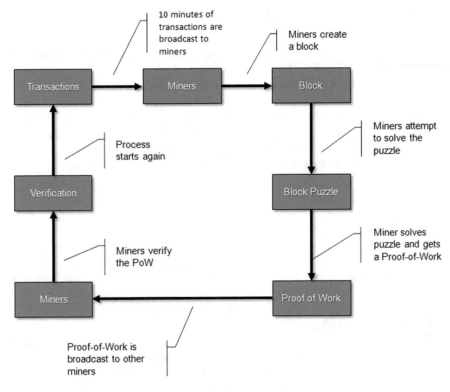

Fig. 4.1 Bitcoin block creation process

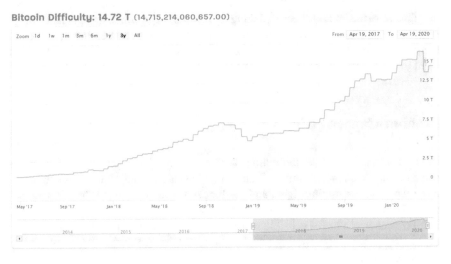

Fig. 4.2 Bitcoin difficulty over time (Retrieved April 19, 2020, https://www.coinwarz.com/mining/bitcoin/difficulty-chart)

Other Consensus Mechanisms/Architectures

Consensus is how the various DLTs determine which blocks get added to the ledger. We have learned a bit about the process by which PoW verifies a block and adds it to the chain. Also, in PoW there is a rule that essentially says, "the longest chain wins". But there are other mechanisms for determining who gets control. In some permissioned DLTs, only certain entities are allowed to create blocks and they might use a round-round approach (every validator takes a turn verifying blocks). In the rest of this chapter, we will take a look at some of the more popular approaches. As a manager, you will want a passing knowledge on these approaches.

Proof-of-Stake

The Proof-of-Stake DLT makes a design trade-off and assumption that the owners of a system will be incentivized to ensure that it works reliably. Recall that PoW was designed, in part, to solve the problem of those pesky, untrustworthy Byzantine Generals. Instead of relying on those mechanisms, let us just give everyone that controls the DLT a stake in the system!

With PoS, there is no competition among miners to solve the puzzle associated with creating a block. Instead, an algorithm picks which peer in the system creates the block based on their "stake" in the system. Thus, those that invested the most have a greater chance of being the entity that solves the block puzzle. This is usually based on how much they have invested in the system. The tokens for PoS are sold in what is known as an "Initial Coin Offering" or (ICO). This is when investors have an opportunity to "stake" the system, determine how much they are going to invest, and thus rewarded with tokens matching their investment. Since there is no mining, there is no reward for creating the block, so instead the peer creating the block is simply awarded a transaction fee. Tokens can still be bought and sold if investors want to increase or decrease their stake.

Proof-of-Capacity (Fig. 4.3)

In this consensus scheme, the mining devices that participate in the DLT network use a mechanism that determines their available space on their hard drives to determine who has the mining rights and the blocks that are validated. In this scheme, instead of solving a complicated math puzzle as is done in a PoW consensus scheme, in a PoC scheme a list of possible solutions is stored in the empty hard drive space of the device before the mining activity starts. The bigger the hard drive, the more potential solutions one can store on it.

Fig. 4.3 Hashing has many useful features – but you don't eat this kind

To get started, a list of all the possible nonce numbers are created by repeatedly rehashing the data, which each nonce containing 8192 hashes. Once the nonce numbers have been created, the PoC mining begins. This is a process where the miner generates a "scoop" number. This tells the miner which part of a nonce to go to. That number will be used to generate a "deadline", which in this case, is the number of seconds that must elapse from the creation of the last block in the chain, before a new block can be created. Deadlines for every nonce on the miner's hard drive are calculated. The nonce with the shortest deadline is selected by the miner. If no other miners can beat this time, then this miner gets to create the block (Fig. 4.4).

One beneficial feature of this scheme is that if you need the computers storage for something else, you can delete as many nonces as you need for the space you require – and keep on mining, albeit, with a lesser chance to "win" because your computer stores fewer nonces.

One benefit for anyone interested in mining PoC consensus-based blockchains is that it is "ASIC proof" (ASIC computers are the special type of computers built for PoW mining – see the Chap. 6). This is because ASICs are designed for numbers of hashes per second – which is CPU-dependent. In these early days, anyone could use storage. However, storage has its own analog to Moore's law wherein the density of data that can be put in a given space continues to rise – and the price of hard drives continue to fall. Were PoC to become more popular, we might see a race for storage capacity much like we see a race for hashing capacity in the PoW domain.

For more information on PoC consensus and how it is used, Burstcoin is the "pioneer" of PoC consensus and provides a mechanism for people to begin mining (if they have some extra unused disk storage available).[3]

---

[3] https://www.burst-coin.org/introduction/for-miners/

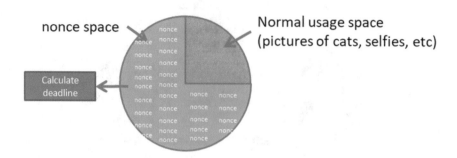

Fig. 4.4 Proof-of-capacity example hard drive

Proof-of-Authority

> The Irony of Trust
>
> Proof-of-Authority – because we don't trust *you*.
>
> Proof-of-Work – because we don't trust *anyone*.

In Proof-of-Authority (PoA) consensus, the validators are chosen ahead of time (although additional validators may be added later). However, the numbers of validators is usually kept to a low number, on the order of a couple dozen or so. The validators are *permissioned* and only known entities are allowed to validate other identities and transactions and are allowed to add blocks to a chain.

Because all the validators are known, and theoretically good actors, all the work that is required to succeed in the presence of bad actors, such as we see with PoW, is not required. Thus, PoA can have very high transaction rates and also, since the validation is not computer-intensive, you do not see the power consumption that occurs with PoW consensus networks.

With higher transaction speeds and lower energy usage (which means less expense to run the operation), why would anyone *not* use PoA.

PoA networks are not as decentralized – so with fewer nodes in the network, all things being equal, it is easier to take down a PoA network, say, with a DDOS attack. Additionally, for those interested in privacy, that does not exist in PoA consensus as each validator is known. But perhaps the biggest is censorship prevention. Recall how in Chap. 1 Nava Ravikant referred to Bitcoin as "political insurance"? In PoA consensus, validators can stifle any speech (by way of the recorded transactions), they do not agree with. They could refuse any transactions or even allow them to participate on the network. Remember, DLT is not just for cryptocurrency, but any smart contract – an exchange that could also be any recording of intellectual property claims, identification, or other transactions that would normally be an exchange of paper, can be recorded for all posterity on a blockchain... unless the entities that control the blockchain do not want to allow it.

Thus, the irony of trust. In PoA networks, the guiding principle is, "we don't trust *you*"; thus we are going to restrict access. But in PoW the guiding principle is, "we don't trust *anyone*"; thus anyone can participate and not have their transactions (speech) restricted.

Directed Acyclic Graph

Directed Acyclic Graph (DAG) is a different type of DLT. IOTA[4] is the leading DAG-based platform, based on the work described in "The Tangle" white paper.[5] (There are other DAG-based DLTs including Nano and OByte. Hashgraph also employs DAG for time-sequencing transactions). In the IOTA DAG, each transaction that is added approves two prior transactions in the tangle as shown in the figure below. Out on the "edge" of the figure, these unapproved transactions are referred to as "tips".[6]

There are different strategies for how transactions are approved. "Lazy tips" refer to new transactions that point to older transactions. Old transactions have already been approved so these lazy tips are not helping with the development of the tangle. An algorithm was developed so that an incentive would be created to reduce (but not eliminate) lazy tips based on the cumulative weight of the prior transactions (Fig. 4.5).

How are IOTAs created? They are not mined like we saw in the PoW consensus mechanism, but rather all the IOTAs were created in the genesis block (transaction

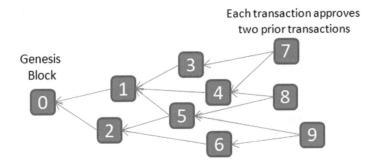

Fig. 4.5 A conceptual example of the IOTA "tangle"

[4] https://www.iota.org/

[5] https://coincompare.eu/wp-content/uploads/2018/10/IOTA-Tangle-whitepaper-at-CoinCompare.pdf

[6] https://blog.iota.org/the-tangle-an-illustrated-introduction-4d5eae6fe8d4/

0 below). The IOTAs were then distributed to investors in the IOTA project. Thus, one function of an approver is to verify that if two parties exchange IOTA, then the spending party has the IOTA to begin with. To avoid "double spending", negative balances are not allowed in the IOTA chain. Thus, not only does the approver need to make sure that the prior transaction has the balance to cover the transaction being approved, it needs to make sure that no other spending is occurring that would make the balance go negative. If this transaction was approved, then the approver's own transaction would not be approved because it was based on a prior bad transaction to the network.

"There is a market need for distributed and fast consensus without the need to centralize the consensus process. ... Traditional blockchains rely on storage of all events across all members of the network"[7] The blockchain model facilitates the ability to easily query for values, but as we have seen in the earlier discussion, this limits the speed at which transactions can be handled.

Semidot Infotech did a great job in breaking down the benefits of Hashgraph over PoW blockchain noting:[8]

- Up to 50,000 times faster
- Level playing field – transactions are handled in an order based on their timestamps
- Fast verification – noting that there is no need to carry prior transactions so the data requirements are under 1 GB (as of this writing the Bitcoin blockchain is at 317 GB.)
- Hashcash is much more efficient because it does not rely on the costly (from an energy perspective) processing that PoW uses.

A conceptual model of a hashgraph is shown in the figure below. In this network, there are four nodes, A, B, C, and D. Events occur through time, and as each occurs it is handled based on their timestamps. The events also link to each other. When a new consensus cycles occurs, the nodes check to see if they agree on the events that occurred in the previous cycle. For this "agreement" to work, the nodes must be connected to the previous cycles' events.

Proof-of-Cooperation (Fig. 4.6)

Proof-of-Cooperation was created in 2017.[9] Developed by Thomas König, this PoC uses "cooperatively validate nodes (CVN)" to create blocks in its blockchain. In the pdf[10] used to describe PoC, the authors state, "Both mechanisms [mining and minting] are widely used, but consume a lot of energy and advantage the rich and thus

---

[7] https://hedera.com/hh-consensus-service-whitepaper.pdf

[8] https://semidotinfotech.com/blog/hashgraph-vs-blockchain-making-blockchain-obsolete/

[9] https://wiki.p2pfoundation.net/Proof_of_Cooperation

[10] https://fair-coin.org/sites/default/files/FairCoin2_whitepaper_V1.2.pdf

Fig. 4.6 Conceptual
hashgraph

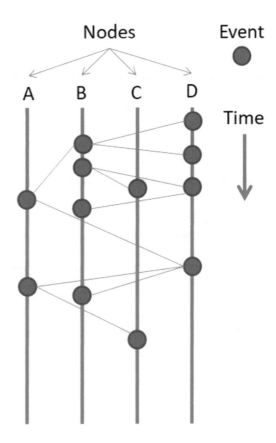

can not be considered as fair". This is the stated driver for switching from a PoS consensus to the new PoC.

Similar to Bitcoin, Faircoin (which is the token the authors created based on this PoC consensus), has a fixed amount of coins (albeit Bitcoin has yet to have all its coins mined). The authors state that no new additional Faircoins will be added. The fixed amount of coins tends to make the coins more valuable over time. Instead of creating new coins, the authors state that, "FairCoin will help to create the conditions for existing coins to be redistributed to amazing social projects worldwide, thanks to funds like Global South Fund, Commons Fund, Technological Infrastructure Fund and Refugees Fund".

In the Faircoin PoC, blocks are created every 3 minutes, using a round-robin method to determine which node will create the next block in the chain. There is an algorithm to determine which validator goes next in the sequence, and checking to ensure enough signatures have been received from the other validators in the network to proceed. While the control of the blocks is distributed amongst the CVNs, anyone can run a node to validate the transactions that are stored on the chain.

There is a second, community aspect behind the creation of Faircoin. The authors wanted to use it as a means to support the FairCoop Ecosystem.[11] Their goal is "building a post-capitalist society that consists of the creation of cooperative relationships in all aspects of life; such as economy, politics, ecology, culture, and human needs." The Faircoin community is not based solely around the cryptocurrency but also this cooperative advocacy and post-capitalist view of the world that they hold. It is not just a cryptocurrency, it is an agenda.

Federated Byzantine Agreement (Fig. 4.7)

Federated Byzantine Agreement (FBA) is a tweak on the traditional Byzantine General consensus problem solving seen in PoW. In FBA, instead of nodes being known and verified ahead of time, nodes do not have to be known and verified ahead of time. Membership is open, (permissionless) and control is decentralized. But the key facet is that nodes can decide what nodes that they want to trust. (If you knew you had a "General" that kept screwing you over, you would stop trusting them.) Thus, consensus is made by the quorums that emerge from the collection of individual nodes.[12] Ripple (using token XRP[13]) was the first to use FBA and as of this

Fig. 4.7 FBA – Pick which nodes to trust

---

[11] https://wiki.fair.coop/en:faircoin:start

[12] https://towardsdatascience.com/federated-byzantine-agreement-24ec57bf36e0

[13] https://ripple.com/

writing is the third largest cryptocurrency in terms of market cap.[14] This mechanism was later improved upon by the Stellar blockchain. From Stellar.org:[15]

> Stellar makes it possible to create, send and trade digital representations of all forms of money—dollars, pesos, Bitcoin, pretty much anything. It's designed so all the world's financial systems can work together on a single network.

FBAs also use a mechanism known as a "quorum slice". That is, they do not need to rely on the entire set of nodes, but can use a subset of nodes that they decide to trust. Thus, if they have decided to trust nodes, x, y, and z, and x, y, and z, are all in agreement, then the local node will also agree.

What happens if they do not agree? These are referred to as "disjoint quorums". These quorum slices that are not connected to other quorum slices could create other quorums which might agree to a different set of transactions, which create a problem for overall consensus.

In the colored grid figure below, these different conceptual nodes have been colored to show which nodes that they trust, with quorum slices colored green, blue, purple, orange, or aqua. However, these are not all disjoint quorum slices as the nodes that are lettered show where the different quorum slices intersect. Green and blue quorum slices intersect at node a; blue and orange quorum slices intersect at node b. Orange and purple quorum slices intersect at node c, and aqua and purples quorum slices intersect at node d (Fig. 4.8).

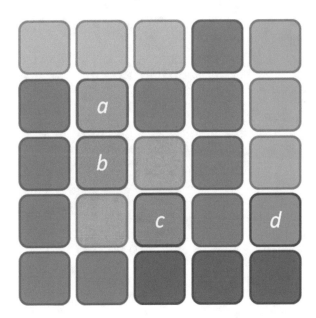

Fig. 4.8 Example of a conceptual federated Byzantine agreement quorum slice configuration

---

[14] https://coinmarketcap.com/

[15] https://www.stellar.org/learn/intro-to-stellar

Questions

1. Bitcoin "halving" refers to the value of Bitcoin being cut in half every four years (True/False).
2. A consensus algorithm that is based on how much available storage capacity you have is referred to as Proof of _____ ?
3. PoA requires each of the participants be _____ ?
4. With Federated Byzantine Agreement, nodes do not have to be known and verified ahead of time (True/False).
5. IOTA is based on the consensus algorithm known as _____ _____ _____.

Chapter 5
Stable Coins and Non-fungible Tokens

Learning Objectives
- Understand the value proposition of stable coins and what makes them "stable"
- Understand what makes a cryptocurrency deflationary versus the inflationary/recessionary aspects of fiat currencies

One of the concerns about various cryptocurrencies is that their value can fluctuate wildly. Recall that this fluctuation was illustrated in previous chapters with charts for both Ethereum and Bitcoin. To address this fluctuation, the concept of a "stable" coin was developed. A stable coin would not be set solely based on the perceived value or demand for a currency, but rather, its value would be pegged to a group of fiat currencies.

Facebook probably drew the most attention in the stable coin space when it announced its "Libra" ($\approx$LBR) cryptocurrency in 2018. The vision for Libra was that it would be enabled by Facebook's Calibra wallet (a subsidiary company) and the customer could use it with the other applications that Facebook owned such as Messenger and WhatsApp.[1] Facebook also put together a coalition of businesses as part of the governance mechanism, with each organization having but one vote. Libra uses an open source blockchain and has its own development language called "Move".

Facebook attempted to head off criticism of Libra by creating the governance mechanism and making Calibra a separate company. Given the criticism of Facebook around data privacy and censorship even before the launch of Libra, there would be concerns that Facebook would be tempted to "spy" on the transactions used to do the same sort of target marketing that the Facebook platform enables. Censorship could also not only include silencing of voices that Facebook does not like to also include banning transactions with parties with which it did not agree. Considering the efforts to "doxx" (to publicly disclose the identity, address, or other personal

---

[1] Welcome to the official White Paper https://www.diem.com/en-us/white-paper/#cover-letter

© Springer Nature Switzerland AG 2021
G. R. Gray, *Blockchain Technology for Managers*,
https://doi.org/10.1007/978-3-030-85716-5_5

details of (someone), especially as a form of online harassment[2]) and deplatform individuals or companies, these concerns appear to have been well-founded. Thus, the governance mechanism ensures that it is not only Facebook that is calling the shots – and as a separate (albeit Facebook-owned company), Calibra would not have access to the privacy data of its users.

The Libra was originally planned to be tied to several historically stable currencies such as the dollar, pound, yen, euro, and the Swiss franc. This approach has since been modified, and the new governing board also has a new name (now named to the Diem Association); it has decided to offer single currency stable coins as well as the multi-currency coin. These single-currency coins include LibraUSD, LibraEUR, LibraGBP, and LibraSGD for dollars, euros, pounds, and Swiss francs, respectively.

The Facebook consortium has probably taken the brunt of the heat relative to stable coins, but what is interesting is that while they were talking over a dozen stable coins were launched. Indeed, after some regulatory scrutiny some of the original members, Visa, Paypal, and Mastercard withdrew, although they were later replaced with Shopify and Tagomi.[3] The table below shows the ten largest stable coins (by market cap), what currency they are tied to, and when they were founded. Some are fairly recent, coming after the Facebook announcement, while others pre-date it by quite a bit (Table 5.1).

If you think about money supply, inflation occurs when there is a surplus of money in the economy. A recession or depression is when monetary supply is reduced. Nominally, the entities that control fiat currencies try to limit inflation while also avoiding a recession. This notion of monetary supply works with cryptocurrencies as well. As discussed in the section on Bitcoin halving, being that it is a limited commodity, like gold, it is deflationary. That is, the value will tend to increase over time. The Tether stable coin found itself in a bit of an inflationary

Table 5.1 10 Largest stable coins (by Market Cap)

Coin	Founded	Set to:	Market cap
Tether	2015	USD	$24B
USD Coin	2018	USD	$4.7B
Binance USD	2019	USD	$1.1B
TrueUSD	2018	USD	$275m
Paxos Standard Token	2018	USD	$244m
Bitshares	2015	USD	$83m
Stasis EURO	2018	Euro	$38m
EOSDT framework	2019	USD	$2.6m
Binance	2019	GBP	$973,119
Stably USD	2019	USD	$521,000

[2] Doxx https://www.thefreedictionary.com/doxx

[3] https://www.cnbc.com/2020/04/16/facebooks-libra-plans-new-crypto-offering-backed-by-just-one-currency.html

predicament when originally it was planned to also have a fixed amount of tokens, but then decided to add more tokens to the supply. This caused the price of Tether to decline[4] back in 2017. Although Tether is pegged to the US dollar, it is not really backed by it. Much like any fiat currency, the value is mostly based on the faith that one places on it.

One can view the market movement of the various stable coins, using the Tether example in the figure below (Fig. 5.1).[5]

The green line shows the price of Tether (very stable), while the market cap is shown with the blue line, and the equivalent price in Bitcoin (BTC), shown in the yellow line. As the green line seems to suggest outside of the "blip" in 2017, Tether has lived up to its class of cryptocurrency and has been very stable.

Why are people using stable coins then? It would appear because it is digital and has lower transaction fees than, say, credit card companies, (3% Visa/Mastercard, 5% American Express), and it is more stable (hence the name) than other cryptocurrencies that are not pegged to fiat currencies. Without the wild fluctuations in price, these sorts of crypto assets become more attractive as mediums of exchange. But we

Tether Chart

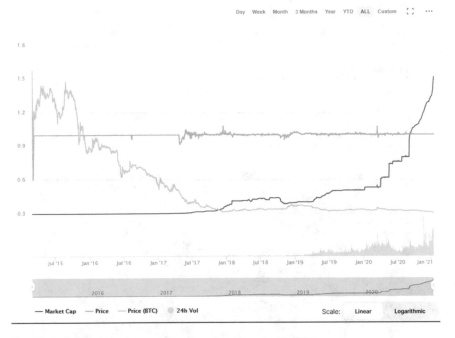

Fig. 5.1 Tether market price comparison, retrieved January 11, 2020

---

[4]Top 13 Stablecoins of 2020 and Their Role on Crypto Market https://changelly.com/blog/best-stablecoins-comparison/

[5]https://coinmarketcap.com/currencies/tether/

are not seeing them used for the creation of distributed applications that is seen with tokens such as Ethereum.

Non-fungible Token (Fig. 5.2)

You may be surprised to learn that "fungible" does not have anything to do with mushrooms. I know I was. It simply means how interchangeable a thing is – if a thing is fungible, it could be swapped out for something of equivalent value or comparable qualities. Non-fungible, then, is the opposite. These are items that are not easily replaced. Non-fungible tokens (NFTs) are specialized tokens that represent

John Pompliano
@JohnPompliano

TikTok sensation Nathan 'Doggface208' Apodaca is selling his viral longboarding video as an NFT.

The starting bid is $500k.

He plans to use the money to buy a home for his parents and fund a new event center in his hometown, Idaho Falls.

Fig. 5.2 NFT digital asset example (Pompliano)

something that is unique. Most are used for in-game assets and collectibles.[6] Some computer games, for example, have unique, digital assets that are intentionally kept rare to increase their value as collectible items – think digital baseball cards. Because the assets are digital, they could easily be replicated – more could be made at any time. But by limiting the number in circulation, the creators hope to create a collectable demand for them based on their rarity. Other types of things can also have NFTs associated with them. "Because NFT are unique, no two are alike, and hence NFT cannot be replaced with another identical token… and it is enforced by smart contracts that prevent duplication, while publicly visible blockchains allow for provable scarcity" [ibid].

In one example of a digital asset being sold as an NFT; Nathan 'Doggface208' Apodaca, is selling a viral video that he created on the TikTok platform. The video showed him skateboarding and casually drinking some juice, while listening to a Fleetwood Mac song. The video became immensely popular – and now being sold via a NFT. It will be a great test of the ability to sell "digital art" via the NFT mechanism.

In another odd "art"-related example, Pringles, the makers of potato chips that stack, which are sold in their uniquely shaped can, created a "limited edition virtual flavor" which was called "CryptoCrisp". Pringles made 50 of the virtual chips with an example being shown on Rarible.com.[7] Rarible is a web site used to buy and sell NFTs secured on a blockchain. The starting price for one of the virtual chips was $2 – roughly what you would pay for a can of Pringles, but at auction bids went as high as $539.

Questions

1. What is the "big deal" with Non-Fungible Tokens?
2. One of the values of stable coins is that they have lower transactions fees than the major credit card companies (True/False)
3. Libra is known as what type of cryptocurrency?
4. The value of a stable coin is tied to what?
5. The value of fiat currency is based on _____?

---

[6] What are Non-Fungible Tokens (NFT)? https://www.ledger.com/academy/what-are-nft
[7] Rarible.com https://rarible.com/token/0xd07dc4262bcdbf85190c01c996b4c06a461d2430:32206
8:0x9b6d720354b24a47ed44d159a77accbb059dcf9e

Chapter 6
Bitcoin Mining and Making Money

Learning Objectives
- The important ratings for a mining computer
- The impact of Moore's law relative to mining
- Potential use cases for businesses whose core business is not mining
- To determine if the high-power consumption of Proof-of-Work processes are a concern

You have learned a bit about how all these cryptocurrencies work and you have learned some of the characteristics of the underlying technology of DLT; now it is time to go make some money, right? Maybe (Fig. 6.1).

Back when Bitcoin first started, if you had a home computer you could set it about the task of crypto mining and have a fairly decent change of successfully earning some mining rewards, and doing this without any specialized equipment. A friend of mine had a son who was an early adopter of mining. He related to me the story of how his son eventually had four computers in his room, all mining. But he had to have a conversation with his son because all he knew for certain was that his electricity bill had gone up significantly. (See a further discussion on mining and energy use in the next chapter.) He pulled his son aside and said he did not think much was going to come of this so it was time to sell those "coins" of yours and these extra computers.

The computers were sold and that investment recovered and the eight coins his son had acquired were sold for… $800. (As of this writing, those coins are worth $287,120.)

Eventually, the landscape of Bitcoin mining changed. The software that ran the mining functions worked better on computers that had high-powered graphics processing units (GPUs). It is the same kind of high-power card that is required for computer gaming. At the time the demand for GPUs were making the gamers unhappy because when you wanted the latest, greatest card for your new computer gaming rig, the miners were snapping them all up as fast as they could be produced.

© Springer Nature Switzerland AG 2021
D. R. Gray, *Blockchain Technology for Managers*,
https://doi.org/10.1007/978-3-030-85716-5_6

Fig. 6.1 I was not always a general

GPUs – just like the CPU that runs your computer, it is influenced by Moore's law.[1] This law is more of a reflection of production improvements and hence is more of an observation rather than say a law of physics. However, it has held up pretty well over the decades with the doubling of circuits that can be crammed into a computer chip being doubled roughly every 18–24 months.

This has not only influenced how much power your computer (and other computing devices, graphics cards, smart phones, tablets, and memory chips) have, but it also influenced the cost as production also scaled up. We are quite gotten used to the notion that the power of computers continues to increase while the relative price of computers continues to decline.

> I think people have a tendency to casually gloss over the significance of computing power increasing this quickly (due to Moore's law). It has enabled the creation of the super computer that you carry around in your pocket. A decade from now computers will likely be not 10 times more powerful than they are today, but 64 times more powerful (assuming Moore's law continues to hold). To be fair, every few years a news headline will garner attention by declaring, "Moore's law is Dead!" And each time the authors of such articles are no doubt disappointed that their prediction has not come true.

Over time miners realized that GPUs were not efficient enough for the mining task. Thus, a new form of computing device was created that specialized in running the mining program. These are called application-specific integrated circuit (ASIC). Your desktop computer and pocket super computer are built for generalized use.

---

[1] Cramming more components onto integrated circuits, https://newsroom.intel.com/wp-content/uploads/sites/11/2018/05/moores-law-electronics.pdf

They are a platform on which programmers can develop many different types of programs. ASICs are built to do one thing and to do it well. If you are only going to focus on mining, you do not need a computer that can also look at pictures of cats on the Internet. You are only interested in earning those tokens. And ASICs can be designed for other single task operations, but the important tasks for this conversation are those designed to do mining.

Vendors have sprung up to meet the demand of computer-specific mining. Let us take a look at a couple of the important specs of the Antminer S9. It is rated at 13.5 TH/S. The "TH" stands for terahash. A terahash is a trillion calculated hashes per second. Thus, an S9 can perform 13.5 trillion hash calculations per second. Not too shabby. And just a couple years ago, this was the standard in performance for mining computers. However, some of the latest models of mining rigs claim up to 110 TH/S as of this writing, almost ten times faster.

One of the other important things to consider is the power rating of the mining computer. The power consumption of this model of miner is 1323 W ±10%, with the suggested power supply. For comparison, a 10,000 BTU air-conditioning unit uses ~850 W of energy.

There is a "computing war" going on in mining scene in the sense that the greater the processing power that one can throw at the hashing problem, the more likely you are to earn a reward. ASIC computers that are specific to mining are continually being updated as new processors are designed thanks to the improvements predicted by Moore's law. If your competitor can earn more coins than you because of greater processing power, this creates an incentive for you to also invest in the latest computing rig in order to not be left behind. Thus, we have seen data centers full of mining rigs attempting to compete at the largest of scales to earn the mining rewards.

How can a single miner compete with that entire scope of computing horsepower? Simple – you cannot. As a single miner, your best bet is to join a mining pool (Fig. 6.2). But joining a pool also has aspects that should be considered. If you join a larger pool operator, the pool is more likely to earn rewards, but that means there is less reward to go around. A smaller pool means a less chance of earning a reward, but fewer people amongst which to spread the reward that is earned. Some additional things to consider are whether the pool operator charges a fee to participate (0.9–4%), what the payout scheme is, how long you have to wait to receive your funds, and how secure and trustworthy the pool is. You would not want all of your hardworking miners to get ripped off in the end.

The table below shows the five largest mining pools,[2] the payout mechanism, and fees (Table 6.1).

It should be noted that these pools are not only for Bitcoin, but for other types of cryptocurrency as well. Bitcoin mining gets a lot of attention, but there are many other types of cryptocurrency that can be mined.

---

[2] https://www.blockchain.com/pools

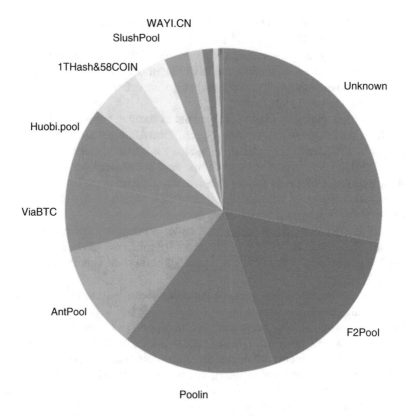

Fig. 6.2 Hashrate distribution among the largest mining pools

Table 6.1 Four largest mining pools

Pool	Payout scheme	Fees
F2Pool	PPS+	2.5%
Poolin	FPPS	2.5%
AntPool	PPLNS/PPS+	0–4%
ViaBTC	PPLNS/PPS	2–4%

When you mine as part of a pool, there needs to be a mechanism to determine who gets paid within the pool, and how much. These are referred to as payout schemes. Let us take a quick look at some of the payout schemes below.

- PPS – Divides your mining power by the total pool miner power; thus you get paid as a percentage of what your mining contributed to the whole.
- Pay Per Last N Share (PPLNS) – In this scheme, miners get paid in a predefined shift. (Shift, as in, when you sent your miners to work – but in this case, when a block has been resolved). The N refers to the number that is the last share to be

solved in a block. If a share is rewarded outside of the shift, you do not get paid. If your miners are working "full time" (your computing power is constant; you did not turn your miner off), the share will be fairly constant. Thus, your shares are proportional to the blocks that were mined during the period.

- PPS+ – PPS+ "is a kind of PPS-PPLNS blend. For each valid share, they submit to the pool, the miners are paid by the pool (regardless of the pool finds a block or not). A portion of the transaction fees that the pool makes is also paid to miners based on a PPLNS method when the pool finds a block".[3]

In terms of mining, the mining itself is its own means to an end. You do the mining, you make money (hopefully), if your rewards are greater than what you pay for the computing equipment and the energy costs. But are there use cases for mining other than pure mining for profit?

Potentially, wherever you have unused computer capacity, when that capacity is not being used, that capacity could be applied to mining. Consider for example, if you have many computers in your office that sit idle in the evening when workers have gone home for the day. This unused capacity could be applied for mining – the same concept applies if you had a data center with unused capacity. Again, the limiting factor is the energy costs for running computers at full speed when there were previous idle. And you would also need to account for any loss of life for computers that would be damaged by excess heat over time. But when you consider that most computers are replaced anyway ~3 years – this may be less of a concern.

Another case in the electric industry is when there is excess energy created for which there is not a matching load to power. This occurs now for example, in territories where the legacy generation (nuclear, hydro, fossil) provides a base generation. With the rise of other renewable energies such as solar and wind, due to the nature of this generation type, it can get really "spiky". During cloudless days or windy days, these kinds of resources generate more electricity. But if the legacy generation is handling most of the load already, a signal is sent to the solar and wind controllers telling them to not send as much electricity onto the grid. (Electric generation has to match the electrical loads or problems happen.) This signal is referred to as "curtailment".

What if instead of curtailing these "green" electrons (green because they are made from renewable sources), you could simply turn on some more load to take advantage? If you had a bank of miners standing by and waiting to be "turned on", they could take advantage of a situation when there is excess electricity being generated.

There are a couple challenges with this idea, however. Computers get replaced ~3 years. The same is usually true for specialized mining computers. Newer, more powerful versions come out so it does not make sense to have these resources sitting idle. The temptation would be to run these computers all the time. But perhaps the case could be made that excess energy could be contracted to be consumed by computers that are already owned and are already idle, in the example of data center computers or office computers that were discussed previously. Perhaps your

---

[3] https://blockoney.com/best-ethereum-mining-pool/

organization might be able to take advantage of such a scheme if mining is not your core business.

Proof-of-Work and Power Consumption (Fig. 6.3)

As Bitcoin became more popular, more and more computing resources were dedicated for the purposes of mining. In the preceding section, we discussed how energy efficiency is a prime concern for mining operators – they want to mine cryptocurrencies at the lowest possible cost. Even with these cost drivers, the energy consumption of miners has continued to grow. But the rise in the energy costs did not really enter the mainstream consciousness until it was pointed out that the combined energy consumption had started to equal that of a small country. Various media outlets pointed to the Bitcoin Energy Consumption Index (BECI)[4] that at the time (2018) showed that the energy consumption was estimated to be 50 TW. Two short years later that estimate has grown to be 77 TWh, which is roughly the energy consumption of the country of Chili. Additionally, the carbon footprint of that energy consumption is estimated to be just shy of 37 Mt, which is roughly the carbon footprint of New Zealand.

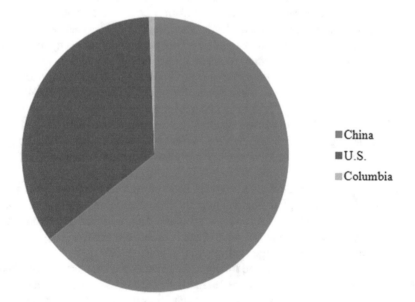

China
U.S.
Columbia

Fig. 6.3 Relative energy consumption of "bitcoin sized" to China/U.S.

---

[4] https://digiconomist.net/bitcoin-energy-consumption

Is that bad? It is difficult to know without context. (Context does not make for alarming headlines so there is less of an incentive to provide it by your local conflict-fueled media organizations.) Columbia ranks 42nd in energy consumption based on 2018 energy estimates from the Energy Information Association.[5] And it is a long way from the consumption used by China (7.2 GWh) or the United States (3.9 GWh).

The other interesting aspect of Bitcoin being singled out for its energy consumption is that there are many other data centers that use a lot of energy. Consider Google or Facebook, or even the U.S. government. How much energy is used so that you can post pictures of your cat or what you are eating? (Facebook uses 5.1 TWh[6] [roughly that of the country of Cameroon] of energy for those of you playing at home and wanting to impress your friends at parties).

Nominally, what energy is used for does not matter. It is all profit and loss. If the energy used for a service provides a benefit to the market, and the value of that service exceeds the expense to generate it, perfect. Capitalism at work – everyone have a nice day.

Now, if you are concerned about where that energy comes from and whether it is renewable – that is another story. If the energy that is used by dirtier sources such as coal, you might be concerned about the carbon that is produced to perform those hashes. But what if the hashes are generated by clean energy? Nuclear, hydro, solar, wind, all generate renewable or carbon-free electricity. Some popular destinations for crypto mining in fact occur close to the source of generation. Areas with a lot of hydro energy in particular have been singled out where miners can place a lot of electric load close to the source of generation, and areas with a lot of hydro-based (dams) electricity tend to have cheaper electricity.

Although if you do have run crypto mining rigs in your house you might be able to use the excess heat to keep your house warm as "Benny" did.[7]

Where else does cryptocurrency mining occur? Not just in the bedroom of my friend, but it is mostly concentrated in locations with low cost of energy and cool climates. The mining rigs generate a lot of heat to run those computers. But unlike traditional data centers that use a lot of HVAC controls to manage the temperature (to prevent premature aging and computer failure), mining rigs are generally run without these sorts of concerns. Often the mining "center" is an old repurposed freight container[8] with a fan mounted on it to circulate outside air through the racks of mining equipment as seen in the figure below (Fig. 6.4).

---

[5] https://www.eia.gov/

[6] https://www.statista.com/statistics/580087/energy-use-of-facebook/

[7] This Is What Happens When Bitcoin Miners Take Over Your Town https://www.politico.com/magazine/story/2018/03/09/bitcoin-mining-energy-prices-smalltown-feature-217230

[8] https://irishtechnews.ie/mobile-crypto-mining-now-available-in-a-shipping-container/

Fig. 6.4 Freight container repurposed for cryptocurrency mining. (Envion AG)

Crypto mining can cause problems for electric infrastructure as well. Consider the situation for most electric utility companies, if you said that you wanted to build a data center in their territory where the computers would be run at full load, 24 × 7, they would take that deal any day of the week. Utilities are in the business of selling electrons. The catch being, if they thought you were going to be around for a while and not just pack up shop if things took a turn for the worse. If you build in a location with preexisting infrastructure (brown field), then the utility does not have to build any new infrastructure to support your operation. But if you build in a new location and then pack up and leave when the cost of energy exceeds the profit from your mining operation, the utility will be stuck with that investment.

In some areas in the United States, mining operations moved into areas that had cheap energy costs, but were also economically depressed; so if the miners move out, there usually are not good prospects for finding a new entity to move, leaving the rest of the community to pick up the tab. Recall our conversation around cryptocurrency volatility. Times are good as of this writing; Bitcoin and others are surging again, but based on the track record, if you were a utility you might be hesitant to make that investment if the mining operation failed.

Salamanca[9] and Plattsburgh, cities in New York state, and Chelan County, Washington, are some of the places that have put a moratorium on crypto mining in their areas due to some of the sorts of concerns outlined here.

Alternatively, the government of Kazuno, a city of approximately 32,000 in Japan, is trying to encourage miners to come to their city, because they have an

---

[9]A City in New York Blocks Mining Operations for Cryptocurrency to Preserve Energy, https://bitcoinexchangeguide.com/new-yorks-salamanca-blocks-crypto-mining-operations-to-preserve-energy/

abundance of renewable energy. Miner Garage Co., Ltd, a cryptocurrency mining company, made an announcement that they would be building a "clean energy mining center" in Kazuno. Without the concerns of CO_2 generation, and the possibility of attracting investment, this city's government went in the opposite direction taken by Salamanca and Plattsburgh.

Another alternative energy source that can be used for cryptocurrency mining has recently emerged. Flared gas – it is the gas that is burned off if producers cannot find a way to process it or if the cost of gas is low and it is not profitable to bring it to market. This gas that is going to be burned anyway can be used to power cryptocurrency mining. It is estimated that approximately 150 billion cubic meters of flared gas is burned annually – roughly the amount of CO_2 produced by Italy.[10] The gas is still burned so it does not alleviate the concerns about CO_2, but it is a more efficient use of gas that would normally be wasted.

Questions

1. A terahash is a measure of what?
2. What are the three main cost considerations for a mining operation?
3. PPS, PPLNS, and PPS+ refer to what?
4. The PoW consensus mechanism uses more energy than other consensus mechanisms. Do you care?
5. What are some of the considerations for a utility allowing crypto mining in their service territory?

---

[10] Flared natural gas latest prize in bitcoin miners' energy quest, https://www.msn.com/en-us/news/technology/flared-natural-gas-latest-prize-in-bitcoin-miners-energy-quest/ar-BB1gMkXk

Part III
Use Cases and Applications

In this section we explore some ways in which DLT can be used. DLT is not a silver bullet – it does not fix or work for every problem. The key is in understanding how the core characteristics of DLT (security, transparency, and immutability) might add value to a new or existing application. Also, as was discussed in the prior section, understanding how the underpinning technologies work and design trade-offs that were made for each type of DLT will also impact decisions about fit of purpose. As a manager, you may be tasked with making fit of purpose decisions about various DLT-based solutions. The goal of this section is to provide some examples, with some high-level assessments based on the DLT core characteristics to give you some ideas about the specific solutions that you may encounter in your role.

Before digging into some of these use cases, we will do a little bit of a deeper dive into the notion of immutability. Following this, we take a look at where DLT is in terms of global investment activity. Then we take a look at distributed applications themselves and where you can go to review the vast array of DLT-based applications. Once this is complete, we will start to dig into some of the specific use cases.

Chapter 7
DLT Core Characteristics of Basis of Use Case Evaluation

Learning Objectives
- Revisit the core characteristics of DLT
- A deeper understanding of immutability
- The trade-offs between better, fast, and cheaper of DLT

If you have been around business and technology for a while, you have probably seen your fair share of "silver" bullets come and go. A silver bullet is a technology that receives a lot of hype and overinflated expectations. DLT – while disruptive to certain types of processes and applications is not applicable to everything. One must consider where DLT might be used based on the value propositions of core characteristics. Also, just because a DLT could be used in an existing process, one must evaluate the capabilities based on how DLT makes a process better, faster, or cheaper.

> There is an old information technology saying that you can have things "better, faster, or cheaper, but you only get to pick two" – you can never have all three options.

Security I can recall when I was regularly doing application development and some manager would say, "We need to design this with security in mind 'from the ground up'". We never did. It was all good intentions. But it was a popular saying. Why? Because you need to make sure the thing works first. You do not want to be troubleshooting security issues when you are trying to get data from point A to point B. You get it to work, and then you add the security controls. When it breaks, well, you know where to look first. But that changed with Bitcoin. Here it actually was designed from the ground up with security in mind. Everything from what hashes are used, to how they are used, digital identities, Merkel trees, and the smart contract itself, security is used every step of the way.

© Springer Nature Switzerland AG 2021
D. R. Gray, *Blockchain Technology for Managers*,
https://doi.org/10.1007/978-3-030-85716-5_7

Transparency This is the characteristic by which every transaction in the log can be verified. You may not be able to see the parties to a transaction, but you can confirm that a transaction occurred and for how much. And if you are a party to the transaction, you can obviously confirm that the terms of the contract were complied with.

Immutability Except as noted where the developer community put a fix in to correct a mechanism by which hackers managed to manipulate an actor in the DLT ecosystem, or where the community has agreed to a feature change, the DLT record does not change. Other than the extreme case where, for example, Ethereum reversed some transactions, even if features are added, what has transpired has transpired, and there is no going back.

Not So Immutable?

What was once thought to be very unlikely has now occurred. Prior to 2020, the hacks of DLT systems had focused on the ecosystem, wallets, and servers that suffer the same sorts of security breaches that other system have suffered. The "51% attack" where an attacker gets control of the nodes and thus can control the contents of the chain itself – and where the concern of double-spending would materialize, happened July 31, 2020, with the Ethereum Classic blockchain.[1] (Fig. 7.1)

Better, Faster, Cheaper (Fig. 7.2)

When considering DLT solutions, keep in mind the old adage "you can have it better, faster, or cheaper, but you can only pick two". Something always has to give. If a solution is "better" meaning it is of greater quality, it usually does not come cheap. If you want a solution faster, chances are you have to pay more to get to a solution faster, or you have to give up some features to get something working sooner. Or you have to scrimp on quality (better). The same holds true for "cheaper". If you want a cheaper solution, you need to be prepared to give up on quality and/or features. As we walk through various use cases later in this section, keep this in mind. The evaluation that is provided here will consider the core characteristics of DLT (security, immutability, and transparency) but will also consider how various solutions impact how a solution makes something better, faster, or cheaper. When vendors approach you with their solutions, these are important questions to ask. If you have a "Greenfield" opportunity (there is no existing capability in your

---

[1] Ethereum Classic 51% Chain Attack July 31, 2020, https://bitquery.io/blog/ethereum-classic-51-chain-attack-july-31-2020

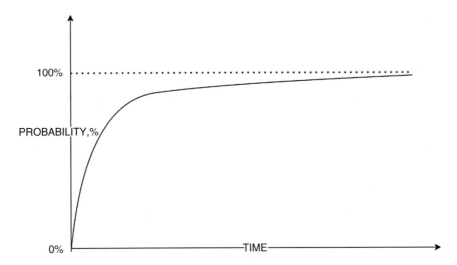

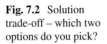

Fig. 7.1 PoW probability theory, which indicates that the level of confidence is very high, but never reaches 100%

Fig. 7.2 Solution trade-off – which two options do you pick?

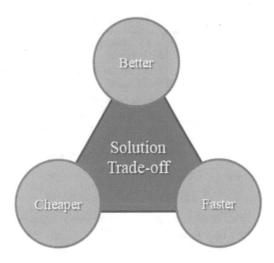

organization), these considerations may be less applicable. However, if you are replacing an existing solution with a DLT-based solution, the value proposition has to have some components better, faster, or cheaper, and you need to understand the trade-offs.

Of course, there are no absolutes in these sorts of trade-offs. You do not get *only* two, but you spread your resources around to make the best of a situation – what is "good enough". Consider the following mental exercise. Consider that you have 10 "priority points" but 15 available slots, would you put five each in two, or would you spread them around. The table below shows an example (Table 7.1).

Table 7.1 Better, faster, cheaper – balanced priorities

Better	Faster	Cheaper
	x	
x	x	x
x	x	x
x	x	x

Table 7.2 Better, faster, cheaper – focus on faster and cheaper

Better	Faster	Cheaper
	x	x
	x	x
x	x	x
x	x	x

Table 7.3 DLT evaluation table

Characteristic	
Security	A brief description of how DLT's security characteristics are applicable to the use case
Immutability	A description of how DLT's immutability characteristics are applicable to the use case
Transparency	A brief description of how DLT's transparency characteristics are applicable to the use case
Relative to status quo	*Opportunity:* A brief description of how DLT might be applied to this use case. *Challenge:* A brief description of the challenges facing a DLT solution for the use case

This reflects a balanced approach – but perhaps your priorities are cheaper and faster. Your table might look like the one below (Table 7.2).

Keep this in mind – there are usually trade-offs for everything, just as there are design trade-offs for the types of consensus algorithms and DLT architectures.

As we walk through the use cases, there will be a discussion about the use case at a high level, and then some consideration of the use case in light of DLT's core characteristics as shown in the example table below (Table 7.3).

Before looking at some specific use cases, let us review where some of the global investment in DLT is occurring, look more closely at distributed applications (dApps), and more broadly at device security.

Questions

1. The core characteristics of DLT are _____ _____ _____
2. Better, faster, or cheaper – which do you choose?
3. Immutable, in the case of DLT, means that the contents of the ledger can never change (True/False)
4. DLT is built with security in mind "from the ground up" (True/False)
5. The ledger behind DLT is "unhackable" (True/False)

Chapter 8
Distributed Applications (dApps)

Learning Objectives
- What is a "killer app"?
- Stakeholders and their respective dApp design considerations

As you have learned, Bitcoin is technically one giant smart contract. But it was not designed to be a platform for other kinds of smart contracts. However, Ethereum, also designed (as of this writing) as a PoW DLT[1] is more suited to this, and hence, is popular as a platform for developing smart contracts, although other platforms and their own dApps have emerged as well. We learned earlier that one of the features of a smart contract is the ability to be automatically executed once the terms are met. However, some dApps work just like other kinds of applications in that they require user input before executing, for example, DLT-based games.

Speaking of smart contracts, let us revisit what a smart contract looks like in relation to other actors that interact with it and with the blockchain. Let us take a look at a sequence diagram that is helpful for viewing these sorts of interactions. If you are not familiar with sequence diagrams, they are a useful tool for documenting interactions. They are read from left-to-right, and from top-to-bottom. Each arrow represents an exchange of information between actors. Thus, an arrow showing a data exchange at the top of the diagram occurs before an exchange that is shown further down in the diagram.

An example is shown in the figure below. This sequence diagram is a high-level depiction of the relationship between the measurement and verification requirements that are needed as part of a transactive energy exchange. Transactive energy facilitates the buying and selling of electricity not just between the utility and their customers, but any willing party. This is just one piece of what a transactive energy system encompasses, but it highlights the interchange between the smart contract and the data that might be stored in a distributed ledger (Fig. 8.1).

---

[1] Ethereum is experimenting with going to a PoS architecture so as to reduce the energy costs associated with running the blockchain.

© Springer Nature Switzerland AG 2021
G. R. Gray, *Blockchain Technology for Managers*,
https://doi.org/10.1007/978-3-030-85716-5_8

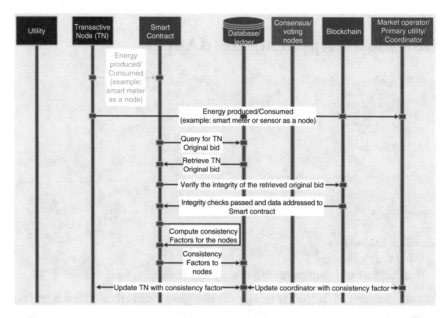

Fig. 8.1 High-level depiction of relationship between measurement and verification require-ments and blockchain features. (Source: Pacific Northwest National Lab (Blockchain Smart Contracts for Transactive Energy Systems, Report August 2019, https://www.osti.gov/servlets/purl/1658380))

It is also helpful to understand some of the basic forms of smart contracts. Florian and Luca[2] do a great job of describing these different types:

- Generic contracts implement application logic, for example, for deposit manage-ment, that can be invoked by blockchain clients or by other contracts; in general, this type of contract is stateful in that it maintains application state across interactions.
- Libraries implement one or more functions, for example, a math library, that are meant for reuse by other contracts; libraries do not store internal variables and are stateless.
- Data contracts provide data storage services inside the blockchain, for example, a client references manager, that are meant for use by other contracts; by design, they are stateful.
- Oracles deliver data services from the outside of the blockchain to the inside of the blockchain, for example, currency conversation rates. Contracts cannot make calls outside the blockchain, as outside dependencies may prevent verifiability (conversion rates change over time). If data from the outside is needed, it can be pushed by clients to oracles using transactions; these then allow other contracts to query for the data.

---

[2]A Service-Oriented Perspective on Blockchain Smart Contracts, https://ieeexplore.ieee.org/document/8598947

"Stateful" simply means that the software keeps track of data that has been exchanged; thus, it is easier for other applications to "keep track" of what has transpired. "Stateless", then, means that the software does not keep track, usually because there is no need.

Returning to our dApps conversation, web sites such as www.dApp.com or www. stateofthedApps.com can view lists and categories of the kinds and popularity of dApps. Each of these sites lists over 1000 different dApps, and the list continues to grow. In fact, chances are that if you name a category, there is probably a dApp for it; insurance, energy, health, finance, games, exchanges, as well as intellectual property protection and other forms of contracts, can all be found on sites such as these.

But one could argue the one "killer app" has not been created yet for DLT. To be sure, adoption of cryptocurrency continues to rise for payments and for currency speculators, but there is not a dApp that is driving the adoption of DLT as a platform by the masses. A killer app is something that as an application makes it a "must buy". A killer application is one that can drive sales and participation. For example, in the early years of desktop computing, business computing capabilities such as spreadsheets eased the lives of business workers, creating the ability to do a variety of business assessments and generating insights drove early computer sales. This was seen with the VisiCalc application for the Apple II series of computers.[3] Roughly, a million copies of the VisiCalc program were sold during its lifetime, followed by more advanced programs such as Lotus 1-2-3, Borland's Quattro Pro, and Microsoft Excel. Today, a million copies of a software release is still considered a pretty good run, but back when VisiCalc first came out, the thousands of computers that were acquired from a need to have the program, was, and still is, considered phenomenal.

Now consider the state of dApps. In the State of the DApps social category, as of this writing, there are between 500 and 1700 daily users of the top five social media dApps. Now, consider a platform such as Facebook whose user number is in billions. Facebook did not get to a billion users overnight of course (Fig. 8.2). (Facebook went from one million users at the end of 2004 to just over a billion users in December of 2012.[4]). There was a time when the number of Facebook users numbered in the 1000s, but it remains to be seen if any social media type dApp will drive that sort of participation. Facebook and Twitter seem to be trying hard to drive users off their respective platforms through various forms of censorship, which may lead to great adoption of a platform that places greater value on freedom of speech. This would benefit dApps because rather than being centrally controlled like Twitter, Instagram, TD Ameritrade, etc., dApps by their very nature use a distributed control mechanism.

---

[3] A Brief History of Spreadsheets, http://www.dssresources.com/history/sshistory.html

[4] Facebook's Remarkable User Growth, https://www.statista.com/chart/870/facebooks-user-growth-since-2004/

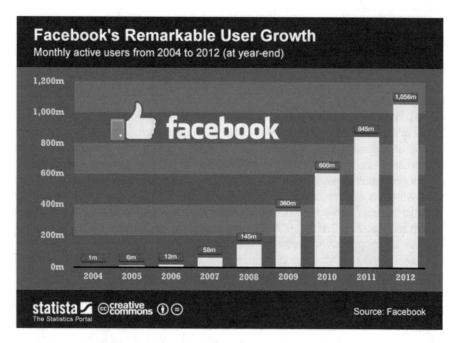

Fig. 8.2 Facebook user growth 2004–2012. (Statista)

Some non-DLT-based social media applications have made the attempt (Gab, Parlour) to compete with Twitter, but have been booted off everything from server hosting to payment services as Silicon Valley attempts to hold hegemony on the social media space.

Although dApps have the benefit of being distributed across a peer-to-peer network such that there is no single point that can be taken down, the dApps themselves could be still be removed from the various app stores that they might be distributed on, which in turn limits the ability to reach a wider audience (Fig. 8.3).

All things being equal, it is harder to buy things that are not in the store (Fig. 8.4).

One implication of social media applications that leverage an immutable ledger is that, well, it is immutable, for good or ill. If a post hurt your feelings, there is no deleting it. Thus, you may have to have a good filter (or a thicker skin if you are easily offended) for platforms that have features such as these. For example, in the Hive FAQ, it is noted that in the PeakD dApp that "content on the HIVE blockchain itself is censorship resistant".[5] As one could imagine, this makes the concept of deletion a bit challenging, again, from the PeakD faq:

Can I delete a post or comment?

- Yes and no. Anything submitted to the HIVE Blockchain is there forever.
- However, another HIVE transaction can change the way it is viewed by interfaces like PeakD including something akin to a deletion.

---

[5] PeakD FAQ, https://peakd.com/about/faq

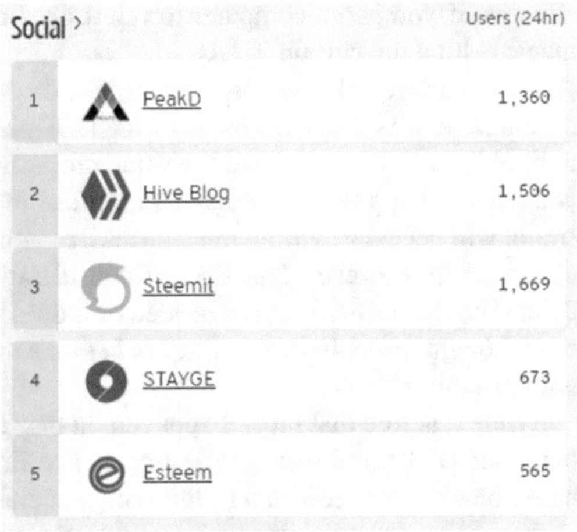

Fig. 8.3 Top five social
dApps per state of the
dApps as of May 4, 2020

Social >			Users (24hr)
1	▲	PeakD	1,360
2	)))	Hive Blog	1,506
3	○	Steemit	1,669
4	◐	STAYGE	673
5	ⓔ	Esteem	565

Fig. 8.4 All things being equal, it is harder to buy things that are not in the store

- Presently PeakD.com does not allow deletions after someone has: A) Commented on your post B) Voted on that post

Source: PeakD FAQ

Although smart contracts are driving disruption across many categories – the disruption is not making users say, "I need to get me some of that DLT", but it might want them to be able to get the dApp that is running on a particular DLT. In the VisiCalc example, the software and the platform that it ran were tightly linked. While you could buy a copy of VisiCalc without the computer, it was *way* more effective if you had a computer to run it on. For many dApps, the platform is the magic behind the curtain.

As a manager, if you consider using a dApp, you may need to understand the design trade-offs that were made of the DLT platform on which the dApp depends, to be able to fully understand the value proposition and design issues, for example, you need a dApp that can close transactions in seconds, not minutes. Such a constraint will necessarily limit the choices that you would want to make. For the *users* of the dApp, however, that kind of consideration will probably be immaterial to them. The dApps performs as needed or it does not. As a manager, you will want to have thought through potential issues *before* a user thinks, "What the heck is wrong with this thing?"

While it is true that most dApps run on the Ethereum platform, some dApps use the other DLT platforms. For example, The Energy Web Foundation has a PoA-based blockchain dedicated to the energy sector. It is designed to be a platform to enable use cases associated with distributed energy resources (DER), such as facilitating the trading of renewable energy certificates (RECs) and device registration. RECs are a means to track the renewable energy generated by various "green energy" resources such as photovoltaic panels and wind turbines. However, being able to track which device generated the energy, and which services it provided, and which organization it provided the services on the behalf of, is an important driver for device registration. Having a platform that stores resource information, its capabilities, its contracts, and associated organizations helps solve "who gets paid at the end of the day".

dApps That Talk to Other dApps

Sometimes dApps need to talk to other dApps to do their work. Sometimes, these will be dApps that run on other platforms, but sometimes these will be dApps that run on the same platform. Sometimes, dApps are run *off-chain*, that is, a dApps may take its processing away from the main chain, often done for performance issues, and then once the computation is completed, then the dApp will re-enter the chain, perhaps posting the results of a transaction to synchronize with the rest of the chain. dApps that run only within the chain are referred to as *on-chain*.

If you need to connect to another dApp or to another platform, it can be useful to use an application designed specifically for this task. The market leader for this coordination is currently Chainlink.[6] Two of its biggest uses are DeFi (a mechanism for integrating price feeds) and Chainlink VRF (a verifiable source of randomness that is used for chain-based gaming applications). These classes of applications are called *oracles*. One example of how this might be used is a smart contract that would

---

[6] Chainlink, https://chain.link/

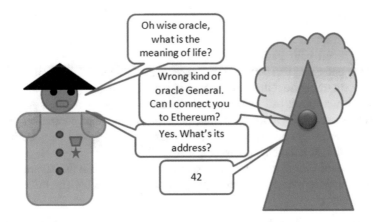

Fig. 8.5 An oracle connects dApps across chains, or between on-chain and off-chain activity

consume price-feed data, and then automatically execute a contract based on the value, perhaps triggering a buy or sell order. Another solution might be to determine the payout, perhaps of that automated sell order, but perhaps the account is listed on a different blockchain. The oracle enables the synchronization of such a transaction (Fig. 8.5).

A new competitor to the Chainlink oracle recently emerged on the IOTA platform.[7] As you probably recall from the discussion in Chap. 4, IOTA is a directed acyclic graph form of DLT. As a DAG-based DLT, the transactions occur much more rapidly than for a PoW-consensus type of DLT.

IOTA boils down their advantages thusly:

- IOTA transactions are feeless
- IOTA transactions can hold a fairly large amount of data
- IOTA's network operates in near real time
- Fetching data using an IOTA node is lightweight and efficient
- IOTA Oracles support diverse security and data structuring capabilities

Source: IOTA oracles[8]

One action taken to improve data quality is that IOTA does not allow data from third-party data providers, but rather, the data must come from first-party sources that write into the IOTA tangle. Thus, there is an incentive that the data is of good quality and accurate.

---

[7] IOTA Blockchain Introduces Oracle Capabilities, https://beincrypto.com/iota-blockchain-introduces-oracle-capabilities/

[8] IOTA oracles, https://blog.iota.org/introducing-iota-oracles/

Bad Contracts – Bad dApps

You may recall your basic Business Law 101 that says roughly that people are allowed to enter into bad contracts, with some restrictions.[9] If you were going to be party to a smart contract that a dApp was using, you would probably want to make sure that it was not a bad one, especially since we know that smart contracts are executed automatically once the terms are met. While smart contracts speed how quickly business can be done, there is no guarantee that it will be *good* business. Caveat emptor, "let the buyer beware", still applies.

Nikolic et al identified a few bad forms of contracts within the Ethereum blockchain that they termed the "greedy, prodigal, and suicidal".[10] The three types are described below:

- Prodigal – A vulnerability where a contract gives away Ether to an arbitrary address which is not an owner, has never deposited Ether in the contract, and has provided no data that is difficult to fabricate by an arbitrary observer
- Suicidal – Contracts often enables a security feature of being "killed" (by issuing a software command called "SUICIDE") by its owner (or trusted addresses) in emergency situations such as when being drained of its token or when malfunctioning. There is vulnerability if a contract can be killed by any arbitrary account.
- Greedy – Contracts that remain alive and lock Ether indefinitely, allowing it be released under no conditions. You can put money it, but you cannot take it out.

Nikolic et al. evaluated 970,898 Ethereum-based smart contracts. In their evaluation of these contracts, they found:

- 1504 prodigal contracts
- 31,201 greedy contracts
- 1495 suicidal contracts

Thus, roughly 3.5% of Ethereum contracts appeared to have some sort of issue associated with them. This probably does not seem like a high number… unless it is your contract. Which reinforces the notion that just because "it's a computer program" does not mean it is infallible, because chances are it was written by a fallible person, so, due diligence is still required.

---

[9] Notable Court Cases Concerning Contracts, https://www.lectlaw.com/files/lws49.htm

[10] Finding The Greedy, Prodigal, and Suicidal Contracts at Scale, https://arxiv.org/pdf/1802.06038.pdf

Questions

1. What is a killer app?
2. "Stateful", when referring to dApps refers to what?
3. What is the ramification of a post being made on DLT-based social media platform?
4. In the context of a dApp – what is an oracle?
5. Some of the bad forms of Ethereum contracts were classified as _____ _____ _____

Chapter 9
Global DLT Activity

Learning Objectives
- Get an understanding of the prevailing sentiments regarding DLT impact on competitive advantage
- Get a sense of the scale of global investment in DLT

With the amount of disruption being caused by DLT, you can be sure that it is drawing attention on the global stage. We explored later in the final section how you never want your sales pitch to be "but blockchain", but in the early days that was often enough to earn a bucket of cash. Investors are probably a little more discerning these days (probably). One can certainly see the amount of attention DLT has just from looking at the market caps of some of the largest cryptocurrencies. The top ten cryptocurrencies by market cap are shown in the table below (Table 9.1).

There are other facets to these cryptocurrencies, such as if they are a currency or a software platform, or if the value is tied to a fiat currency (stable coin), or allowed to change based on demand, and whether the supply is fixed and what the currently available supply of a given asset is.

But will investment continue? In a Deloitte survey, the following were some of the top investment concerns of respondents (in descending order):[1]

- Implementation – replacing or adapting legacy systems
- Sensitive information concerns
- Potential security threats
- Lack of regulatory clarity
- Challenges in forming a consortium
- Lack of in-house skills
- Uncertain ROI
- Lack of compelling application

---

[1] Statista, Barriers to greater investment in blockchain technology worldwide from 2018 to 2020 (Deloitte), https://www.statista.com/statistics/878686/worldwide-investment-barriers-blockchain-technology/

© Springer Nature Switzerland AG 2021
G. R. Gray, *Blockchain Technology for Managers*,
https://doi.org/10.1007/978-3-030-85716-5_9

Table 9.1 Market cap of top ten currencies

Asset	Price	Market cap
Bitcoin	$37.028	$693B
Ethereum	$2,483	$288B
XRP	$0.86	$86B
Cardano	$1.53	$49B
Stellar	$0.33	$35B
Polkadot	$22.87	$24B
Uniswap	$23.60	$22B
Chainlink	$22.97	$22B
USD Coin	$0.99	$22B
Litecoin	$167	$11B

Market Cap of Top Ten Cryptocurrencies, Source: Coindesk.com, retrieved, June 10, 2021

- Unproven technology

Do any of these reasons resonate with you, or have you heard these same kinds of concerns in your organization? If you are creating a product – do you have responses ready to address these concerns?

Disruption in the Utility Industry

Jumping specifically to the energy industry, the global breakdown of investment is shown in the figure below. (IOT is getting a lot of attention across industries.) Transactive energy and decentralized generation and creating a trading platform for those activities is getting a lot of attention, probably due to the nature of DLT potentially being able to provide a platform where others have not been successful previously and the potential revenue from a market that would have a high volume of transactions (Fig. 9.1).

Recall the discussion regarding disruptive technology from Chap. 1 and trying to understand if a new emerging technology is disruptive. The Indigo Advisory Group made an evaluation of the utility industry and made the graphic below to highlight areas where they felt that DLT would be disruptive versus supported and other areas where the impact might be limited (Fig. 9.2).

Now consider your own industry. The key areas that are being disrupted spans industry sectors. In particular, consider the areas highlighted as disrupted in the utility sector, but are probably also being disrupted in other sectors as well. Corporate Strategy, Legal/Compliance, Finance, Supply Chain, and Performance Tracking – these categories of disruption are not limited to the utility industry but will most likely be disrupted in your industry as well.

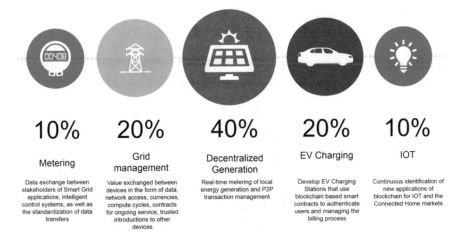

10%	20%	40%	20%	10%
Metering	Grid management	Decentralized Generation	EV Charging	IOT
Data exchange between stakeholders of Smart Grid applications, intelligent control systems, as well as the standardization of data transfers	Value exchanged between devices in the form of data, network access, currencies, compute cycles, contracts for ongoing service, trusted introductions to other devices	Real-time metering of local energy generation and P2P transaction management	Develop EV Charging Stations that use blockchain based smart contracts to authenticate users and managing the billing process	Continuous identification of new applications of blockchain for IOT and the Connected Home markets

Fig. 9.1 Portion of DLT investment in energy. (Source: Indigo Advisory Group (https://www. indigoadvisorygroup.com/blockchain))

But the question lingers… will these levels of investment continue, increase, or decrease? Looking at some developments, one might be tempted to wonder if the bloom had come off the DLT rose. In February of 2021, it came to light that apparently IBM was scaling back its blockchain team, with reorganization and reassignment of resources, realigning focus more towards the hybrid cloud. The revenues expected from blockchain-related development were not realized and IBM saw a decline of revenues by 6% in 2020.[2]

The 2019 Gartner Hype Cycle shows a variety of DLT-related capabilities. Only a handful of capabilities have been headed for what Gartner Research calls the "plateau of productivity". The plateau reflects when the hype has worn off, but useful implementations of a technology have begun to occur (see figure below). This indicates that there are many capabilities that are still headed for peak hype. It may be simply that IBM has not been able to leverage these capabilities as a mechanism for generating revenues – perhaps others are being more successful (Fig. 9.3).

The market report for global blockchain investments included some interesting numbers and observations that reflect how the market may still be growing including:[3]

- Estimating that the blockchain technology market will grow at 62.73% CAGR through 2026 reaching $52.5 billion
- Consortium/hybrid blockchain will be the largest North America blockchain tech area at $6.7 billion by 2026

[2] IBM scales down blockchain activity, https://www.bitreporter.com/blockchain/ibm-scales-down-blockchain-activity/

[3] Global Blockchain Technology Market Report 2021–2026: Accenture will Lead the Charge for Systems Integration and Companies like Amazon, Dell, HPE, and IBM, https://finance.yahoo.com/news/global-blockchain-technology-market-report-092300412.html

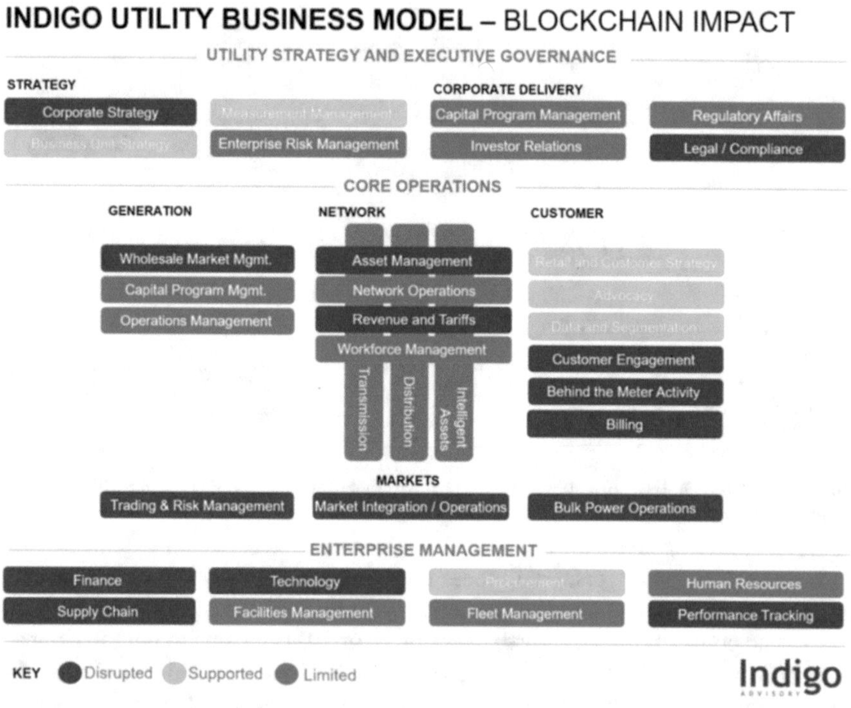

INDIGO UTILITY BUSINESS MODEL – BLOCKCHAIN IMPACT

Fig. 9.2 Utility DLT disruption. (Source: Indigo Advisory Group)

- Anticipate substantial merger and acquisition activity in this space (Fig. 9.4)

Deloitte Consulting also sees an increase in the expectations from participants in their annual blockchain survey. In the 2020 survey, 55% of respondents indicated that blockchain "will be critical and in our top-five strategic priorities[4]", increasing from 43% making the same assessment in 2018 survey. Perhaps even more important was the sentiment that 85% agreed with the statement, "My organization or project will lose a competitive advantage if we don't adopt blockchain technology", up from 77% in the 2018 survey. Although examining just these two response categories, most think that their organization will lose a competitive advantage due to blockchain, and yet almost a third less indicate that it is not a strategic priority – showing a bit of a disconnect between the observation that something will impact their competitive environment, yet are not making a move to do something about it.

Seeing an increase in merger and acquisition activity reflects a growing maturation in a market. When a new innovation appears on the scene, if it looks promising, many existing and new companies look to jump on the bandwagon. This has

---

[4] Deloitte's 2020 Global Blockchain Survey, From Promise to Reality, https://www2.deloitte.com/content/dam/insights/us/articles/6608_2020-global-blockchain-survey/DI_CIR%202020%20global%20blockchain%20survey.pdf

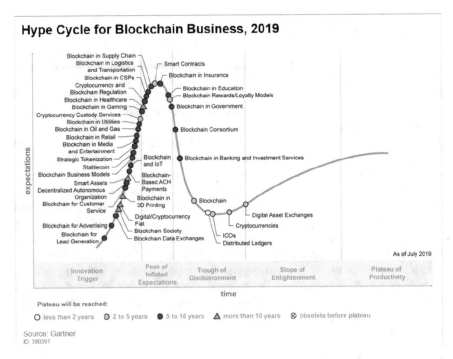

Fig. 9.3 Gartner hype cycle for blockchain business, 2019. (Source: Gartner Research (Gartner 2019 Hype Cycle for Blockchain Business Shows Blockchain Will Have a Transformational Impact across Industries in Five to 10 Years, https://www.gartner.com/en/newsroom/press-releases/2019-09-12-gartner-2019-hype-cycle-for-blockchain-business-shows))

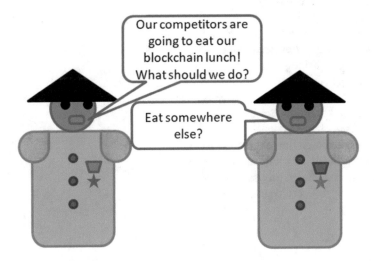

Fig. 9.4 Deloitte survey: 85% indicated that "my organization will lose a competitive advantage if we do not adopt blockchain"

significantly occurred in the DLT space. The next evolution occurs when weaker entrants either go out of business, or are acquired for their promising implementations. Bigger tech companies, if they cannot compete with startups, will often simply buy them to acquire their technology. We are probably many years away from the next phase of market maturation, where you begin to see oligopoly among a few larger players. But this merger and acquisition phase suggests we are well along the maturation cycle for DLT, only cementing the notion that we are indeed looking at a very disruptive technology.

In the United States, the National Science Foundation continues to entertain DLT-related proposals. As part of their Industrial Innovation and Partnerships (IIP), and the SBIR – Small Business Innovation Research/STTR – Small Business Technology Transfer offers seed round funding for small businesses. The IIP "invests in high-tech small businesses and collaborations between academia and industry to transform discoveries into innovative commercial technologies with societal benefits.[5]" There are two phases: Phase I which is focused on feasibility research of a 6–12 month timeframe of up to $256 k, and Phase II research where the awardees focus on creating prototypes (24 month timeframe with awards up to $1,000,000). In the latest round of evaluations, these are the categories of proposals that were reviewed. Not all of these proposals were selected; these are shown here simply to give the reader some ideas about what some are thinking of in terms of DLT development:

- E-commerce/shopping
- Customer data management
- Data management/dissemination
- Privacy governance
- Reducing price volatility
- Food provenance
- Project management improvement
- Parts provenance/reverse engineering
- Reputation tracking
- Device management
- Improving logistics
- Labor market improvements
- Improving claims process
- Real estate prediction engine
- Self-healing distributed ledger
- Agricultural market process improvement
- Improving self-service processes

---

[5] www.nsf.gov/eng/iip

Questions

1. What is your bet – will investment in DLT increase, decrease, or stay about the same? Why do you think so?
2. What are some of the drivers of value of cryptocurrencies?
3. What are some of the main concerns of managers regarding DLT?
4. What activity tends to reflect a greater maturation in a market?
5. By market cap – can you name three of the top ten cryptocurrencies? (No peeking!)

Chapter 10
Industrial Internet of Things (IIoT) and DLT

Learning Objectives
- The Internet of Things promises many wonderful things – but for many applications the hype often exceeds reality
- Understanding some of the strengths and limitations of distributed computing devices

It is thought that DLT will bring significant benefits to the Industrial Internet of Things (IIoT) domain. This could be especially important if the devices have the smartness to be able to work and communicate together in a distributed setting, without human intervention. But what is IIoT?

Simply put – think of how the Internet today connects people. Web browsers use Internet protocols to share information between people. (Although the Internet pre-dates web browsers, being used to exchange files and email long before it became important to share pictures of food and cats). The Internet as we know it started as a Defense Advanced Research Projects Agency (DARPA) research project and was used to connect just a handful of locations. The TCP/IP protocols are used to route information – with the primary driver being that data should be able to go around trouble spots (e.g., if a city turned into a giant smoking nuclear crater, you might want to still be able to get messages around that sort of disruption). Then of course, along came the web browser in 1993, then the browser wars between Internet Explorer and Netscape navigator and the advances of the hypertext markup language (HTML) standards etc. all in an effort to connect humans.

Then on the scene arrived a plethora of devices, not used to share data between humans, but to share information between humans and devices. Smart phones, smart meters, all kinds of industrial sensors, all using the same Internet technologies to collect and transmit data such that humans could talk to each other better, but also to provide greater situational awareness of what was happening on the factory floor or out on the electric, gas, or water grids.

But what if devices could be connected and interact without human intervention? That is the opportunity. After all, humans are rather slow-thinking beasts. Consider

© Springer Nature Switzerland AG 2021
G. R. Gray, *Blockchain Technology for Managers*,
https://doi.org/10.1007/978-3-030-85716-5_10

the SENSE-TRANSMIT-DECIDE-TRANSMIT situation that occurs, for example, when operators are controlling the electrical grid. A device senses that something is awry, and then sends the data associated with that event to a system operator. The system operator then examines the data, considers it in light of other situational data that they are receiving, and then sends a control message back out to the sensor. Also consider that there is latency in time that takes a message to be transmitted over whatever network it may be on.

Conversely, the same scenario, except that when the sensor senses something is awry, it has the capacity to take an action, or to work with other sensors or devices in its local area, to take action. The response is reduced to milliseconds. Now, the device should still send a message back to the central control because the system operator will have broader situational awareness, in this example, of the entire grid. The operator can still send command and control messages to the device – this part of the scenario remains the same, but sensors and other devices are networked in this "Internet of Things" rather than of humans.

IIOT and Inflated Expectations

Devices that are being deployed are getting more and more powerful. However, there is a difference between the capabilities of these types of devices, and say, the supercomputer you have in your pocket (smart phone), or laptops and desktop computers (Fig. 10.1). Vendors are trying to balance capabilities, useful life, and cost. A great example of these trade-offs are seen in the deployment of smart meters, which were also one of the first large-scale IIoT networks. By 2018, there were almost 90 million smart meters deployed in the United States.[1] Smart meters have been replacing the old electromechanical meters, and in addition to having the ability to count electrons (usage), also have communications built in so that they can be read automatically, and in some cases, sent signals to update their security, check diagnostics, or send automatic disconnect signals.

A smart meter is essentially a low-power computer that specializes in counting electrons (or water, gas, steam, etc). Now try this thought experiment. Imagine that

Fig. 10.1 Caution: Inflated expectations ahead!

---

[1] https://www.eia.gov/tools/faqs/faq.php?id=108&t=3.

you have to go on the Internet or you peruse a local big box store to buy a new computer. Ah, but there is a catch. You do not get to update this computer whenever you want, you cannot replace it like you do a smart phone (once every couple of years). No. You have to choose a computer that will work reliably and securely for the next 10–15 years, ideally without being touched. Why? Every time a meter fails, someone has to drive out to check on it, and/or replace it. The cost to drive to and from a meter location and replace a meter exceeds the cost of the meter! The business case for deploying a meter, value adds such as automating the readings (instead of having humans do it), and getting automated outage detection or great power quality awareness, is quickly destroyed if the meters constantly need to be touched. We cannot have people driving around rebooting them if something goes awry. This is also why the firmware (the software inside the meter) needs to be able to be updated remotely, for example, if a new security patch needs to be applied, because again, requiring humans to load them by hand and reboot them is cost prohibitive.

Now consider your pocket supercomputer. Chances are the retail cost of that item is hundreds of dollars. It can do an amazing number of things. But you do not want to stick that on the side of a house to count electrons. That is significantly more horsepower than is required for the job.

Why discuss this? To set expectations about the kinds of devices that get deployed in an IIoT setting, their computational capabilities are based on where they are deployed and the types of jobs that they are designed to do, and the design of the devices that require that they be cheap to deploy, yet also secure and reliable. No small task.

Now consider the computational requirements for devices that would operate as a mining node or peer on a DLT network. To put this into context, let us consider the capabilities if a typical electrical meter, of which millions are deployed, and a Raspberry Pi. If you are not familiar, a Raspberry Pi is one of makes of super low-cost computers (Fig. 10.2).

The comparison can be seen in the table below.

Note that the meter's CPU is less than half as fast as the Pi's, and has a tenth of the dynamic storage, but the meter has a significantly greater amount of long-term storage (Table 10.1).

Fig. 10.2 Raspberry Pi Model 3 (www. raspberrypi.org)

Table 10.1 Comparison of Raspberry Pi to an electric meter

	Electric meter	Raspberry Pi 3
CPU	500 Mhz, 32 bit	1.2 Ghz 64 bit
Memory	128 MB RAM	1 GB RAM
Storage	128 GB Flash	8 GB Flash

A meter is built for one job. For electric meters, that job is counting electrons (aka measuring consumption). These types of devices needs to last for 10–15 years, and be able to store measurement records for 30–40 days in case something goes wrong with the communications. It is built to do this job at the lowest cost possible. The Raspberry Pi is also built to do its job at the lowest cost possible, but its job is to be a full-functioning computer so it comes with several USB ports to enable extensions, as well as a camera and an HDMI port for video.

And yet... can the Raspberry Pi be a mining node? Not really. It lacks the storage. Although, technically, its storage can be expanded via the USB ports, using a USB connection does not fit within the standard Raspberry Pi case, so a custom case would need to be created to support such an arrangement. It terms of processing power, it is underpowered for that task. An Electric Power Research Institute study[2] found that the "village" (six Raspberry Pis representing six loads that transacted electricity pricing amongst them using a Ethereum private blockchain) had to use this private blockchain because it could not hold the chain containing all of the prior transactions. Now, a Raspberry Pi has more than enough computing horsepower to run wallet software – so a Raspberry Pi could certainly be the basis for transacting via a wallet and participating in a blockchain or other DLT (Fig. 10.3).

In a distributed world, one way that this could be accomplished is by using a federated network (Fig. 10.4).

Each of the nodes, n1–n10, could transact via a wallet, with one of the mining nodes represented by A and B. The A and B nodes would be one of the peers on the distributed network, bundling up transactions into blocks for addition into the blockchain. Although this is being discussed in the context of the distributed network for devices, this is exactly what happens in the blockchain as a whole. Different wallets are constantly generating transactions and bundled up into blocks by the mining peers. The only difference is in realizing the distributed devices such as those in a distributed grid such as that in an IIoT have to manage expectations of what the devices can and should be doing. The nodes need to be low cost, reliable,

---

[2] Blockchain Market Simulation Using Ethereum: "Blockchain Village", https://www.epri.com/#/pages/product/000000003002015175/?lang=en-US.

Fig. 10.3 Generic
distributed network

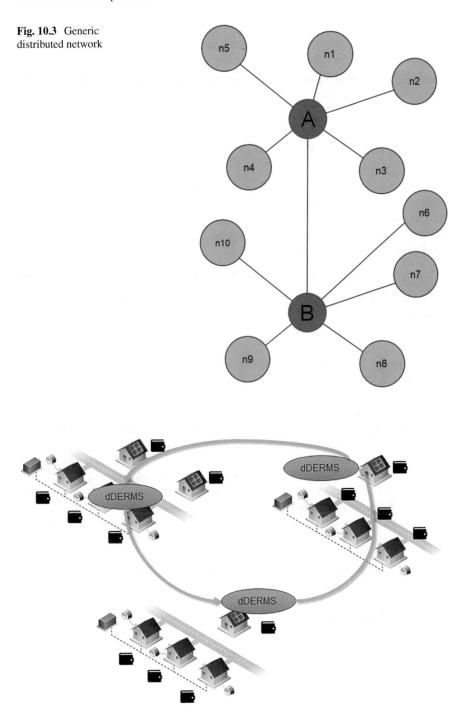

Fig. 10.4 A conceptual model of federated mining peers and wallet nodes

and have long lifetimes (unless like the smart phone model – you can convince someone else to replace it every two years), and mining peers will have shorter time spans (to be effective being replaced with every new generation of mining equipment), but more processing power, although in a distributed grid these would most likely also be remotely managed and not in a data center.

Securing IIoT

We could have billions of IIoT devices, and if they all use DLT to communicate we would not need to worry about security! Not so fast. Although a device may use DLT to communicate, there are still a few things to consider. In this section we will talk about the 50% node problem, securing the device versus securing the messages, and botnets and why they are a problem.

> Even if your devices are not being used for mining, you still need to be able to trust their communications and that they have not been compromised.

What is the 50% node problem? As we learned in the previous sections, each DLT uses a consensus mechanism to validate when blocks of transaction are added to the chain. In a PoW scheme such as Bitcoin, this is really a simple majority. Whatever entity controls the majority of the nodes determines the content of the blockchain. Obviously then, you would not want a single entity controlling more than 50% of the nodes. As of this writing, the top three miners of Bitcoin were approximately, F2Pool, 19%; BTC.com, 14%, and AntPool, 11%.[3] If a single entity were to gain control of the majority of the nodes, the controlling entity could potentially prevent blocks from being verified or allow "double spending" to occur. However, if any entity became too close to this controlling majority, the community would make this an issue as the blockchain at that point would become untrustworthy. But this assumes that the change would be gradual – and hence, noticeable.

What if – and this has not happened, but what if there was a massive takeover at scale of numerous devices – what then? Well first, hacking a device is different than hacking a mining pool. But we do know of one case where hacking millions of devices, albeit small, led to the largest distributed denial of service (DDoS) attack. This was the Mirai botnet.[4]

---

[3] https://www.blockchain.com/charts/pools.

[4] https://www.csoonline.com/article/3258748/the-mirai-botnet-explained-how-teen-scammers-and-cctv-cameras-almost-brought-down-the-internet.html.

What is a Botnet?

A botnet is a group of devices whose security has been compromised, and the controlling entity uses this army of devices for attacks such as a DDoS. A denial of service attack is where the attacker attempts to overwhelm a server with more requests than it can handle. When the server becomes overwhelmed, for all intents it appears to be offline and cannot handle any requests. With a distributed attack, the server cannot block a single IP address from which the attacks occur as with a single attacker. With large botnets the attack could span millions of IP addresses. With the Mirai botnet, the hackers compromised small devices, for example, home security cameras. Typically, the cameras, or other devices, never had their default security settings changed, so they were easily compromised, with the owners of the devices being none the wiser. This is one case where although the individual devices might not have much processing power, they could easily contribute to a DDoS attack.

Why bring this up in light of IIoT? Because even if your devices are not being used for mining, you still need to be able to trust their communications and that they have not been compromised. We have seen with DLT the messaging infrastructure is secure. The use of signed keys and encrypted traffic ensure that this is so. But if the device itself is compromised, via physical or other hacking, the messages emanating from such a device may not be those authorized by the device owner. This is the difference between securing the message versus securing the device.

There have been advances in IIoT device security, by ensuring the devices require that the default security configuration be changed the moment the device is installed. Another improvement is only allowing command and control message to the device be allowed from "white listed" IP addresses. That is, a list of safe IP addresses – if a message comes from an IP address that is not on the list, it is ignored. This greatly reduces the chance that the device will respond to nefarious commands. Although a message might be sent from a white-listed IP address, one must assume that this device has also not been compromised.

Another thing to consider is the physicalsecurity of the device. In the picture seen below, this "smart lock" requires a fingerprint, or just the right set of screwdrivers to open (Fig. 10.5 and Table 10.2).[5]

While the above discussion was related to generic IIoT, the following table will assess DLT fit specific to one form of IIoT: metering. Nonelectric meters (water, steam, gas) need an energy supply for communications, normally in the form of a battery; thus, they typically only send messages a couple times a day. While electric meters may only send message two or three times a day, they can be configured to send messages with greater frequency without concern for draining a battery.

---

[5] https://twitter.com/LockPickingLwyr/status/1007613178249965569.

Fig. 10.5 Not so smart lock.
(Credit: @LockPickingLwyr)

Table 10.2 DLT fit: IIoT

Characteristic	
Security	DLT would provide for greater security mechanisms for IIoT communications
Immutability	DLT would provide an unchangeable record of transactions. Physical device security would be required to ensure that what was written to the log is what was intended
Transparency	The ledger would provide an open log of all transactions, command and control actions, measurement, and device configuration changes.
Relative to status quo	IIoT is a rapidly changing environment, with many new kids of devices and sensors constantly being developed and deployed. *Opportunity:* DLT would provide greater security capabilities than what exist with many IIoT devices today. *Challenge:* The computational capability of the device may limit what DLT capabilities it can use. Adding additional capabilities will drive up device cost.

Electric meters are "always on", and thus have greater configuration options for communications. Electric meter networks are also some of the largest IIoT networks that have been deployed today, with ~100 million smart meters having been deployed in the United States alone. However, electric meters and their communications networks are typically legacy systems, with meters only being replaced every 10–15 years.

Questions

1. What is a 51% hack?
2. Is it cost-effective to manually reboot a meter (or other distributed device?)
3. What is a botnet?
4. Do distributed devices such as those used in IIoT have the computational capability to run a DLT node?
5. Another name for your pocket supercomputer is _____ _____?

Chapter 11
E-Voting

Learning Objectives
- The electronic and physical issues that need to be addressed with electronic voting machines
- The characteristics that one would expect a trustworthy electronic voting system to provide

Do you remember "hanging chads"? During the 2000 national election, the Electoral College results hinged on votes being counted in the state of Florida. One of the big controversies of that election was the result of these hanging chads. During the recount, reviewers examined the ballots that were used, a card type that was designed to be read by machine. The voter would punch a hole in the card to indicate their selection. The card would be then put into a machine to tally the result. However, some chads were not cleanly punched, and some merely "dented" by the device used to remove the chad (the chad being the tab of paper that would be removed by punching a hole in the card). Thus, those charged with hand-counting the results had to determine the intent of the voter. Not the best situation. And in the 2000 election, one with consequences went all the way to the United States Supreme Court.[1] Surely there must be a better way, and that better way would seem to be e-voting.

Blockchain-based e-voting has been tried on a limited basis – but it shows promise for what could be. In their article, *Blockchain-Enabled E-Voting*[2] the authors noted that a blockchain-enabled system had been implemented by the city of Moscow and has more than two million users. City officials had the security of the system audited by Price-Waterhouse-Coopers (PwC) and found "no reason to be concerned". Although for my money, I would feel more comfortable had the system been subjected to a black hat hackathon. If *hackers* say a system is secure, I tend to believe that more than when a paid consultant says something is secure.

---

[1] How the 2000 Election Came Down to a Supreme Court Decision, https://www.history.com/news/2000-election-bush-gore-votes-supreme-court.

[2] Blockchain-Enabled E-Voting https://ieeexplore.ieee.org/document/8405627.

© Springer Nature Switzerland AG 2021
G. R. Gray, *Blockchain Technology for Managers*,
https://doi.org/10.1007/978-3-030-85716-5_11

DLT seems to be well-suited for e-voting solutions. The built-in security and the immutable ledger address gaps in prior failed e-voting solutions.[3] Using a voting app on a smart phone or other computing device, the voting would spend their token to participate in the election (or other voting event) or their choice. In addition to using a cryptographic identity, the devices and users should implement two-factor authentication and biometric identification to ensure that the cryptographic identity matches the device users' identity.

There are challenges to be overcome for a DLT-based voting system. Consider the figure below. With elections that use vote counting equipment today (shown generically as "counter" in the figure below), a person fills out their ballot and the ballot is run through the reader to tally their input. This is where the systems in use today have challenges (Fig. 11.1). For example, the hanging chad issue that was discussed at the opening of this chapter. It *seemed* like a great solution. The user punches holes in a card and then the counter reads the card. Simple, right? After all, the designers must have thought, "computers have been reading punched cards since the dawn of the computer age, what could go wrong?"

Modern counting machines still read paper ballots, but instead of punching holes in them voters fill in boxes. And even though the instructions indicate that things like checkmarks and x's will not be counted – people still do it. And yet people that cannot follow these simple instructions are still allowed to vote. Whattayagonnado? Another challenge came to light in the United States 2020 election in that some ballots were filled in with Sharpies, not pens. (Sharpies are a brand of permanent markers.[4]) This is probably due to the ink bleeding through the ballot and confusing the

Fig. 11.1 Conceptual e-voting DLT architecture

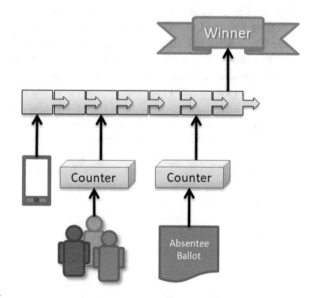

---

[3] Virginia Finally Drops America's 'Worst Voting Machines' https://www.wired.com/2015/08/virginia-finally-drops-americas-worst-voting-machines/.

[4] https://www.sharpie.com/.

counter. Then, just as with the hanging chad situation, other humans have to determine the intent of the voter. This "human in the loop" is one of the inherent flaws in any system. One would assume good faith attempts at getting the vote right – but humans have a bad habit of acting oddly when their feelings get in the way. To combat this, observers are added to the equation – and you would hope that kind of oversight would be all that is required, and yet we have seen situations where observers were interfered with in various ways.

These challenges are broadly lumped into data quality issues. There is an old saying with computer systems that "Garbage in = Garbage out". That is, if bad data is allowed in the system, the result will also be bad. In an election, you hope that at least the result is directionally accurate if it cannot be specifically accurate. And to be sure, all systems have data quality issues – and the more impactful bad quality data is, the more is spent to correct for it. And yet, election integrity with electronic voting machines still eludes us. If you want to have faith in your election, there seems that there should be a better way.

What would you want from an e-voting system?

- Convenience – you would want to encourage participation
- One person – one vote
- Accuracy – you would want to know that the vote that you cast was the vote that was tabulated
- Confirmation – you would want to be able to confirm that the vote that you entered was the vote that was counted
- Security – you would want a system that was secure from external influence or local tampering

What are some ways that DLT-based e-voting can help – and some things to consider that DLT-based e-voting will not help.

One way that DLT-based voting could potentially help as shown in conceptual architecture figure is with smart device-enabled voting. Smart devices often have biometric-based security features that can be used to confirm that you are who you say you are. Additionally, there are other mechanisms that could be employed with two-factor authentication, such as indicating your desire to vote, and having the voting system notify your device when it is ready for your vote – thus, you could have a mechanism to verify who you are with the voting system, tied to a device that allows you to vote – but only when you have initiated the interaction. This is the heart of a two-factor scheme – something you have (smart device) and something you know (an identity that can be confirmed). If such a mechanism was employed, it would certainly be convenient – you would not even need to take time off from work. The authentication mechanism would be used to confirm that you only voted once – and if another person attempted to vote after you vote was tabulated it could be flagged (and if someone had attempted to co-opt your identity, this could also be flagged). With a digital system like this, a voter could receive an electronic receipt confirm their selections. Also, using a DLT-based encryption, the message traffic from the smart device to the ledger registering the vote would be secured, and of course, as we have explored earlier in this book, the ledger itself is secure.

How will DLT-based voting not help – there remain a few challenges. One must assume that not everyone will have a smart device and be capable of voting via one. For this, the legacy challenges remain. You would still need a mechanism for tallying the votes at a counter. Any electronic ballot counter still relying on reading a paper ballot will have the potential for error. What may improve is by connecting the counting device to the ledger tallying the votes by a secure DLT-based security mechanism.

Nominally, e-voting machines are not supposed to be connected to the Internet to avoid remote hacking of the devices. However, as was seen with the United States 2020 elections, at times this prohibition was disregarded (either intentionally or unintentionally). Better to have a connection – *but only* – to the ledger that is tabulating the results. Computing devices can have what are known as "white lists" of allowable connections to and from the device. The same DLT-based encryption could also be used to secure the electronic traffic between the counting device and the ledger. There would need to be an additional proviso: that the software running on the counter is open source and available for inspection. The e-voting systems that are in use today are not open source. Open source software is generally considered more secure than its non-open source counterparts. This is because with open source software, anyone can view the code – and when errors or issues are spotted, they can be corrected. Sometimes, some people attempted to employ a "security by obscurity" approach – using non-standard software and non-standard communication mechanisms in the hope that it will thwart potential attackers. The challenge is that this approach usually leads to software that is not appropriately maintained (a lack of available skills). And if a hacker decides to devote themselves to attacking such a system, should they succeed, it is harder to repair the issue, again, because of fewer people that know such a system.

Physical security of the counting devices and physical security of the ballots themselves would still be a concern with a DLT-based voting system that relied on legacy systems that were connected to them. However, one issue that could be improved with DLT-based voting is the provenance of the ballots (or collection of ballots). Just like with the logistical use cases, the ballots themselves represent a logistical issue. When ballots are moved from smaller polling stations that do not have their own ballot-counting machines, the chain of custody, including who handed off ballots to whom, and when, could also be tracked by the DLT ledger. Implementing a system where ballots were tracked – much like your packages are tracked via UPS, FedEx, or the USPS, would let everyone know where ballots were at in their journey, and who had them. Tracking the movement of ballots via a ledger would improve transparency – that would be open to inspection by anyone wishing to see such information. This makes it more difficult to initiate "shenanigans" or have any surprises about where and when groups of ballots were at – and why they were at certain locations at certain times. Thus, if a group of ballots were diverted, or if a third party attempted to introduce ballots that were not from an authorized polling station,these would be quickly identified.

Electronic security of legacy counting devices would still be an issue – just as we have seen with the hacks associated with computing systems and devices that are in the ecosystem of the various ledgers. White listing devices so that they receive only authorized electronic traffic from known entities helps – but it does not help if the devices themselves have had other software on them comprised by a hack. For example, we saw this with the ledger hack when the customer relationship management system was hacked – exposing customer data. And though it was not a hack of a ledger, but rather a hack of numerous systems that used the SolarWinds monitoring software, it has implications for electronic counting devices that are connected to the Internet and how even trusted sources can be compromised. SolarWinds is a company that provides monitoring software that many companies use to remotely keep track of computers and what is happening with them. In the SolarWinds hack, this software was compromised with malware. The hackers also used Amazon Web Services cloud hosting to disguise their intrusions as benign network traffic.[5] Could such software be used to compromise electronic voting machines? Well, it was found to have been installed on some Dominion Voting Systems – although it is unknown if the monitoring software was used to compromise these voting machines. Although one cannot anticipate when a trusted software provider might be compromised, it does beg for the use of DLT for configuration management.

Configuration management software – really a use case itself is software that keeps track and logs changes to a computer or other device. With DLT-based configuration management, changes that are made to software components, and data such as their related version and date of installation, could be written to an immutable ledger, with open audit ability for any interested party management. Thus, any changes are tracked cannot be tampered with. With an immutable log of transaction, it is clear what was changed, what the change was, and who initiated the change. If there is a change, for example, with an electronic voting system, if there is a last second change, the details and timing of the change are transparent.[6]

Also, if there are "glitches" that for some reason change votes[7] raising questions about electronic voting systems and these systems in particular, these sorts of issues can be tracked down when the votes are tallied in DLT ledger – but they cannot fix human-in-the-loop issues, either malignant or benign if the equipment itself is being misused or ballots being tampered with. But these sorts of DLT-based improvements could certainly help with the status quo and help restore confidence in the voting mechanisms employed today. And, potentially, one day these legacy systems with their human-in-the-loop issues will be eliminated (Table 11.1).

[5] SolarWinds not the only company used to hack targets, tech execs say at hearing, https://www.cnet.com/news/solarwinds-not-the-only-company-used-to-hack-targets-tech-execs-say-at-hearing/.

[6] Fix coming to Georgia touchscreens to restore missing Senate candidates https://www.ajc.com/politics/fix-upcoming-to-georgia-touchscreens-to-restore-missing-senate-candidates/ASEWAGDAR5DFPGW2OPULV4N3JY/.

[7] Dominion Voting Systems Unwrapped https://letsfixstuff.org/2020/11/dominion-voting-systems-unwrapped/.

Table 11.1 DLT fit: E-voting

Characteristic	
Security	DLT's built-in cryptographic "keys as identities" provides security and confidentiality for voters. DLT's security provisions could also be used to protect against other forms of attacks by securing the communications channel – white listing allowable domains and using the built-in electronic signatures.
Immutability	DLT's nominally immutable ledger would provide an unchangeable record of transactions and a means for the voter to verify that their vote has not been altered. It could also provide a means to determine how the vote was recorded, when, and by what authority.
Transparency	The open log of all transactions could be audited to ensure only eligible votes were cast and that eligible voters only voted once.
Relative to status quo	*Opportunity:* A DLT-based e-voting machine might finally create a reliable system that users could count on to be secure. The ability to cast a secure vote from anywhere using a smart phone or other device would make voting convenient which may incentivize participation. Open source smart contracts could conceivably be a strong competitor to incumbent closed-source e-voting systems used today. *Challenge:* E-voting has had many notable failures with hackers being able to demonstrate how the voting machines have been vulnerable to exploits. Moving e-voting to smart phone or other computing device would require safeguards to ensure that these devices are not compromised.

Questions

1. Whose opinion do you trust more – an auditing company, or a hacker, when it comes to the software system security?
2. What is a "white list"?
3. You can always trust the third party vendors that provide add-on software to your system (True/False)
4. What is the job of configuration management software?
5. If you have electronic security, you do not have to worry about physical security (True/False)

Chapter 12
Financial

Learning Objectives
- Some of the ways in which financial institutions have been testing DLT to improve the speed of transactions.

DLT is poised to disrupt the fax machine? Yes. Amazing to think that in this modern information age, many financial transactions are still done with paper and fax machines. This was a bit of a surprise to me when I bought our last house ten years ago. "You want me to fax what? What's a fax?" Of course I knew what a fax was, I was just incredulous that I would be expected to transact with one – or even have one at my disposal.

You can see this theme repeated in Martin Arnold's article – *Five ways banks are using blockchain.*[1] The five ways Martin identified are:

- Clearing and settlement – Martin cites an Accenture[2] estimate that $10 billion could be saved by automating the processes associated with clearing and settlement.
- Payments – Banks, as we noted earlier in this book, are part of the DLT ecosystem that may be associated with exchanges to convert from fiat currency to crypto and back again – and they may set up their own cryptocurrencies for the purpose of automating payments.
- Trade Finance – Bills of lading and letters of credit, sent around the world by fax
- Identity – "Regulators hold banks responsible for checking that customers are not criminals or illicit actors, and fine them if they get it wrong" – and to anyone that has suffered from identity theft, you know how important this protection is

---

[1] Five ways banks are using blockchain, https://www.ft.com/content/615b3bd8-97a9-11e7-a652-cde3f882dd7b.

[2] Banking on Blockchain, A Value Analysis for Investment Banks, https://www.accenture.com/t20170120T074124Z__w__/us-en/_acnmedia/Accenture/Conversion-Assets/DotCom/Documents/Global/PDF/Consulting/Accenture-Banking-on-Blockchain.pdf#zoom=50.

© Springer Nature Switzerland AG 2021
G. R. Gray, *Blockchain Technology for Managers*,
https://doi.org/10.1007/978-3-030-85716-5_12

- Syndicated loans – "A syndicated loan, also known as a syndicated bank facility, is financing offered by a group of lenders—referred to as a syndicate—who work together to provide funds for a single borrower. The borrower can be a corporation, a large project, or a sovereign government. The loan can involve a fixed amount of funds, a credit line, or a combination of the two".[3]

You see the words "paper" and "fax" appear again and again. To be fair, identity is not a core financial use case; as you have read in these pages, the authentication and protection of identity is repeated throughout many of these use cases. But it is clear that in the financial space that was never a single platform that "won". So organizations pass around paper and faxes to transfer data between systems and people. Part of this was due to needing a signature for some documents and transactions. This is moderately surprising given that electronic signatures were authorized for most transactions, yet paper-based signatures stubbornly refused to go away.

Although fiat currencies are where the traditional finance organizations have made their money and have their expertise, they have also been investing in DLT start-ups or developing their own proof-of-concepts to expand their service offerings. However, there is somewhat of a dichotomy in regard to privacy, authentication, and transparency in financial DLT. The big banks have created their own PoA DLT. PoA, as you recall, requires that participants are required to be identified. You cannot be anonymous on a PoA-based DLT. In this paradigm, to be trusted you must be known. And yet, firms participating in such a DLT normally do not like the fact that due to the transparent nature of DLT others can have access to the details contained with the ledgers' record of transactions (Fig. 12.1).

These sorts of details have implications for parties to the transactions. For example, knowing how much a customer has been for an item could lead to more accurately predicting how much a customer might pay for it or related items. (Some would no doubt be horrified that I do not buy anything "on sale". When I need a thing I just go buy it.) No doubt Amazon and Paypal probably know similar things about me – because although they do not use DLT they know all of my transaction history. A leader in the hotel industry, Marriot's position is in part due to their leadership in big data analytic capabilities. Marriott gathers mountains of data and can leverage their analytics for competitive advantage. They developed dynamic pricing automation which they use to adjust price on a variety of factors such as weather, demand, reservation behavior, average daily room rate, among others.[4] Marriott gains these insights from data gathered from their own systems – they do not need to pull data in from a DLT ledger, although technology-savvy; if the data was readily available, they would no doubt use as just another data stream to feed their algorithms (Fig. 12.2).

As we quoted Naval Ravikant earlier in this book, Bitcoin (and other cryptocurrencies) is political insurance. One criticism of cryptocurrencies is a concern that

---

[3] Syndicated Loan, https://www.investopedia.com/terms/s/syndicatedloan.asp.

[4] Why is Marriott the Big Data analytics leader in hospitality? [Case], https://blog.datumize.com/big-data-analytics-in-hospitality-marriott-international-case-study.

Fig. 12.1 Blockchain has the potential to disrupt the FAX as a means of closing financial transactions

Fig. 12.2 Our old friends...

this is just another bubble – people hold up the example of the tulip speculation. If you are not familiar, it was one of the biggest speculations of all time in the 1600s centered in Holland. "At the height of the market, the rarest tulip bulbs traded for as much as six times the average person's annual salary".[5] Like many market failures,

---

[5] Dutch Tulip Bulb Market Bubble Definition, https://www.investopedia.com/terms/d/dutch_tulip_bulb_market_bubble.asp.

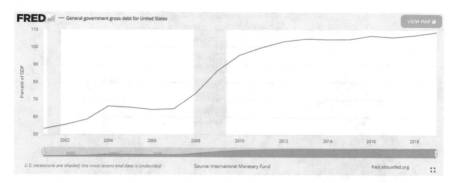

Fig. 12.3 Debt of the United States as a percentage of GDP. (Source: Federal Reserve Bank of St. Louis, with data from the International Monetary Fund (General government gross debt for United States (GGGDTAUSA188N), https://fred.stlouisfed.org/series/GGGDTAUSA188N, retrieved March 20, 2021))

people assumed that the price would continue to climb, so if they took loans to buy tulips expecting the price to go up but when demand waned and the price went down people were forced into bankruptcy so the cycle went negative as quickly as it had arisen. One of the drivers of speculation is faith: Faith in a commodity – or currency. And we have certainly seen a fair bit of speculation in cryptocurrency. But is fiat currency any better? Fiat currencies are no longer back by commodities such as silver or gold. They are merely back by the word of the government that there is value. But that value is only as good as the faith the people have in the currency, and by proxy the government that is backing it. When a government such as the United States takes on an incredible amount of debt and where the revenues brought in go to mostly service that debt, and as the value of the dollar declines, it is probably no wonder that faith in the dollar may also be waning. As seen in the figure below, the debt of the United States is now greater than its Gross Domestic Product (GDP). If the United States was an individual lender, you probably would not have much faith in its ability to pay back its debt (Fig. 12.3).

In the longer term, it may be that a cryptocurrency, such as Bitcoin, has a finite amount of availability – there are no new "bulbs" that can be planted, giving it dis-inflationary properties, will come to have more trust in it than fiat currencies.

Finance-Related Proof-of-Concepts

A Japanese firm, Mizuho Financial Group, announced that they were going to develop a financial payment platform for IoT devices.[6] The idea is that the bank would offer mechanism for payment beyond what you have with mobile phone or

---

[6] Japanese bank Mizuho begins development of IoT payments platform, https://internetofbusiness. com/mizuho-iot-payments-platform/.

Table 12.1 DLT fit: Financial

Characteristic	
Security	DLT-based technologies such as digital signatures can be used to verify transaction participants are who they say they are.
Immutability	Using DLT not only can increase the speed at which contracts are executed, a complete, electronic, and immutable audit trail for every transaction that a financial institution made would improve the accuracy of audits required by third parties. (For more on this, see Chap. 14 Auditing – What DLT was made for)
Transparency	Financial institutions are big fans of transparency… for other people and institutions. DLT levels the playing field, for good or ill. With financial regulations that require banking participants to be identified – this provides assurances for the banks that they are dealing with known entities.
Relative to status quo	*Opportunity:* Anything that replaces a fax machine is probably a good thing. DLT-based identity management would nominally make identity theft harder. Using digital IDs would require criminals to bypass the added protections of things like two-factor authentication, and digital signatures and changes to identity would be captured in their immutable ledger. *Challenge:* There is an entire ecosystem of people "with their hands out" relative to significant transactions. Consider the act of buying a house; appraisals, inspections, title insurance - every step of the way requires payment for something to someone. That is a lot of smart contracts that need to be written.

tablet computers. Mizuho is partnering with SORACOM, a Japanese IoT platform provider. Together, they will jointly develop the SIM cards and circuit boards these devices will use to connect to their network. One of the first applications of the partnership is to see if fingerprint sensors on the IoT devices can be used to initiate bank transfers and to check the balance of the users' accounts. Mizuho had already partnered with IBM to develop a token-based blockchain settlement system,[7] desiring to learn "how payments can be instantaneously swapped". Combining the lessons learned from using IoT devices to initiate transaction coupled with DLT-based transaction mechanism could give Mizuho (and other financial systems like them) insights into how these technologies work together. As one of the earlier movers in this space, Mizuho may gain a competitive advantage relative to other companies that enter DLT-based finance later (Table 12.1).

---

[7] Mizuho bank partners IBM to test blockchain payments system, http://www.coinfox.info/ news/5782-mizuho-bank-partners-ibm-to-test-blockchain-payments-system.

Questions

1. The value of most fiat currencies is based on the faith that people place on them (True/False)
2. You can remain anonymous in a PoA consensus-based system (True/False)
3. What are some of the categories of ways in which financial institutions are leveraging DLT?
4. What are some of the implications of knowing the entire transaction history of a customer?
5. What is one key difference between a tulip bulb and a Bitcoin relative to availability?

Chapter 13
Logistics

Learning Objectives
- An understanding of some of the technologies that are employed today for the tracking, trading, and shipping of commodities
- An understanding of some of the benefits that have been realized by logistics companies and their trading partners employing DLT-based platforms and smart contracts to improve the transaction tracking and speed

DLT has the potential to be very beneficial in the logistics space. Anywhere, there are paper-based systems that are prone to error, losing information, or "non-delivery losses",[1] DLT can help. This is another domain where this problem could have been solved without DLT, but it has not been. Different systems and vendors often are used in different jurisdictions and the transfer of information between them can be problematic. The Electronic Data Interchange (EDI) was developed, in part, to solve these very issues. EDI is the computer-to-computer exchange of business documents in a standard electronic format between business partners. Instead of paper invoices and purchase requisitions, everything can be handled electronically. However, there are still "human-in-the-loop" issues with these systems when it comes to tracking the goods associated with a purchase requisition.

Technology such as scanners and quick response (QR) codes has been developed to automated facets of tracking of assets. And as smartphones have advanced in their capabilities, the ability to read QR codes has become a standard function. But this often still relies on human operators to do the scanning. The same challenge applies to radio frequency identification (RFID). An RFID tag can be attached to an asset – it essentially has a small antenna in it. When the tag comes within proximity of the radio reader, it is scanned automatically. In part, a way to remove human in the loop from the process – when the good is transferred the radio scanners could be put in the entry and exit points. But RFID is not perfect either. Consider the case when you have a pallet of assets. The radio scanner can usually read the assets on

---

[1] A euphemism for people stealing your goods.

© Springer Nature Switzerland AG 2021
G. R. Gray, *Blockchain Technology for Managers*,
https://doi.org/10.1007/978-3-030-85716-5_13

the outside portion of the stack, but it cannot read the tags of the assets in the middle of the stack.

How does DLT help this situation? For those that still rely on paper-based systems or processes, it is a good excuse to transition from these antiquated methods of handling transactions. DLT could potentially provide a mechanism for many parties to store their transactional data in a single blockchain. EDI even provides the data model to start so that designing the smart contracts for logistics is not needed to start "from scratch". It still does not solve the human-in-the-loop problems when humans are relied upon to do the scanning. Humans may not remember to scan an asset when it is moved. How often do we "forget" when we are in a hurry – for example in an emergency situation?" But with DLT where an asset "disappeared" can be tracked to the last known location in an immutable ledger that cannot be altered, which helps to prevent "non-delivery losses" (Figs. 13.1 and 13.2).

In one example of "lost" trade transactions, in 2014, after an investigation, China had determined that $10 billion in falsified trade transactions had occurred. "Fake trade has increased pressure of hot money inflows and provides illegal channels for

Fig. 13.1 Example of a
QR code

Fig. 13.2 DLT helps with provenance of goods and eliminates "misdirected shipments"

criminals to move capital in and out of the country".[2] With some banks failing to "fulfill their duties to check the authenticity of trade documents" might cause one to speculate as to whether this was by negligence or due to some incentive being provided to the banking authorities.

In another example of using DLT to ensure that payments reach its intended target without being "diverted" by middlemen, a bitcoin-based startup, Bankymoon, makes it possible for donors to send money directly to South African school meters.[3] In many areas of the world, prepaid meters are used instead of the more common – read the meter, get billed and then pay. Thus, a prepaid meter in many ways acts as an electronic cash register. When the balance in the meter runs out, the power is turned off (unless some sort of a credit mechanism is enabled). Thus, donors can validate their donations reach the intended recipient and can see how much energy their donation has provided for.

In logistics for food supply, IBM has created the IBM Food Trust™, a network of collaborating "growers, processors, wholesalers, distributors, manufacturers, retailers, and others, enhancing visibility and accountability across the food supply chain".[4] The participants use the permissioned blockchain platform to keep a record of transaction data, providing provenance for the source and chain of custody for food products, including a "farm-to-store" view. The platform includes the capability for participants to manage, upload, or edit documents facilitating information management along the supply chain. With its platform, IBM provides four main value propositions:

- Food freshness: Accurately judge remaining shelf life
- Food fraud: Help eliminate the chance for fraud and errors
- Sustainability: Help ensure the promised quality
- Food waste: Help minimize waste hot spots

Walmart, with revenues exceeding $559 billion in 2020,[5] is one of the world's largest retailers and has one of the premier logistics operations. However, Frank Yiannis noted the logistical challenges relating to food safety, stating, "Piecing together traceability data by sifting through hundreds or even thousands of documents during a food borne outbreak can be slow and complicated, and it all too often is not an effective way of identifying and informing action through lessons learned to prevent future outbreaks".[6] Provenance, the ability to trace products from source to destination is not just to ensure proper payments or to prevent the waylaying of shipments, but from a safety perspective, to also being able to quickly track down when a contamination has occurred and to know where the products have been

---

[2] China uncovers $10b worth of falsified trade, https://www.globaltimes.cn/content/883512.shtml.

[3] How Bitcoin Brought Electricity to a South African School, https://www.coindesk.com/south-african-primary-school-blockchain.

[4] IBM Food Trust, https://www.ibm.com/products/food-trust.

[5] Walmart company facts, https://corporate.walmart.com/newsroom/company-facts.

[6] A New Era of Food Transparency Powered By Blockchain, https://www.mitpressjournals.org/doi/pdf/10.1162/inov_a_00266.

distributed so that they can be quickly rounded up to prevent further illnesses. Yiannis noted that in the *E.coli* outbreak case in 2017, it took the Centers for Disease Control (CDC) that after two months, no definitive sources had been found, indicating that between the outbreak in 2006 to the outbreak in 2017, no apparent improvements in food tracking had been made. The 2017 outbreak led to 25 people being sick, cases spread across 15 states, and one death.[7] Yiannis also noted that more frequently recalls were not limited to a single food type such as romaine lettuce that was the source of the *E.coli* outbreaks, but now include specific ingredients in food, making the traceability challenge all the greater. Food fraud is also a concern with examples such as olive oil being "cut" with regular vegetable oil, or dehydrated milk products being cut with fillers; such fraud costs are estimated to cost the industry between $10 and $15 billion annually.[8]

Walmart worked with IBM to develop some proof-of-concept applications to determine the provenance of mangoes from farms in Mexico where they were grown to the store where they were sold. The improvement in the time that it took to find this data improved from 7 days to 2.2 seconds. Of course, it is a greater challenge to go from a proof-of-concept to incorporating the entire product catalog.

In another logistics example, in 2017, a pilot trading implementation was done with participation from ING and Societe Generale, where they tested a solution which they dubbed "East Trading Connect". The exchanges involved transactions from these banks and a Swiss commodities trading company, Mercuria involving an African crude oil shipment. The shipment was then processed by ChemChina. The results were that the pilot reduced the time for the parties to play their roles in the exchange from three hours to twenty-five minutes and led to a cost savings of 30% over how the exchange was previously handled. Bank representatives state that, "the solution delivered on all expected fronts by eliminating documentary fraud, reducing costs, and improving efficiency".[9]

The ability for DLT to provide a platform to better automate the exchanges, especially those associated with international trade, are highlighted by the example from 2014, when the Danish shipping company (Maersk – which handles roughly 20% of the container market) tracked a shipment from East Africa to Europe, "A single container to handle a shipment of goods required stamps and approvals from 30 people such as those in customs, tax officials, and health authorities".[10] DLT may not eliminate all of the conversations and communications that people might have, but even putting a small dent in the number of "humans in the loop" should provide dramatic improvements in the speed of how the transactions are processed.

[7] Multistate Outbreak of Shiga toxin-producing Escherichia coli O157:H7 Infections Linked to Leafy Greens (Final Update), https://www.cdc.gov/ecoli/2017/o157h7-12-17/index.html.

[8] Food Fraud and "Economically Motivated Adulteration" of Food and Food Ingredients, https://fas.org/sgp/crs/misc/R43358.pdf.

[9] ING and SocGen partner on blockchain solution, https://www.businessinsider.com/ing-and-socgen-partner-on-blockchain-solution-2017-2.

[10] IBM And Maersk Apply Blockchain To Container Shipping, https://www.forbes.com/sites/tomgroenfeldt/2017/03/05/ibm-and-maersk-apply-blockchain-to-container-shipping/?sh=216f6fb3f05e.

Characteristic	
Security	Built-in security helps assure that only authorized individuals have access to the transactions that are part of a logistics-based smart contract
Immutability	The immutable ledger can be used to confirm entry and exit points and trace an asset from creation, to delivery, to use, to retirement or disposal – with no means of changing the records by any party.
Transparency	The ledger provides a record of every transaction that can be inspected.
Relative to status quo	*Opportunity:* For those that still rely on paper-based systems or processes, this may be an opportunity to these error-prone methods – although it still relies on humans in the loop for some portions of the system (scanning). If a logistics system has already been digitized or automated, the benefits may be limited to being able to track the transactions in an immutable ledger, eliminating fraud. *Challenge:* There may not be as good of a value proposition for systems that already use EDI to automate exchanges and have automated scanning capabilities such as QR or RFID.

Questions

1. What is the value of provenance for goods?
2. Describe RFID?
3. What is a non-delivery loss?
4. DLT could eliminate all human-in-the-loop bottlenecks (True/False)
5. Why has not EDI solved the problems with logistics provenance?

Chapter 14
Peer-to-Peer Energy Trading and Transactive Energy

Learning Objectives
- Understand the basics of peer-to-peer energy trading
- Understand how a DERMS may play a part in a transactive energy architecture

Transactive energy is the notion that electricity exchanges evolve from being limited to utility to consumer, to one where transactions can occur between any willing participants. In most places around the world, if you generated excess electricity, for example, via rooftop solar panels or by a wind turbine, you can only sell that energy back to the utility for which you are a customer. With the transactive energy "prosumer" concept, anyone could buy or sell energy from any willing participant. Thus, with transactive energy, nominally, if you had a solar panel on your house, and you made extra electrons, you could sell those electrons to your neighbor that did not have a solar panel on their house, yet desired to use more renewable energy. If you know anything about physics, you know it is probably a little hard to bundle up electrons and deliver them to a specific place. Electricity is consumed as soon as it is created, and there are other factors that determine how electricity flows on the grid. What gets traded in practice are renewable energy certificates (RECs) – a document that verifies the renewable energy generation at a location. These used to be (and in some cases, still are) paper-based, and were also limited to 1 MW in total size before being traded. But more and more, these are being converted to electronic format, and of course, using DLT platforms, which now facilitates the trading of smaller quantities. But first, let us take a look at the definition of transactive energy.

The GridWise Architecture Council defines transactive energy as:

"The term 'transactive energy' is used here to refer to techniques for managing the generation, consumption or flow of electric power within an electric power system through the use of economic or market based constructs while considering grid reliability constraints.

© Springer Nature Switzerland AG 2021
G. R. Gray, *Blockchain Technology for Managers*,
https://doi.org/10.1007/978-3-030-85716-5_14

The term 'transactive' comes from considering that decisions are made based on a value. These decisions may be analogous to or literally economic transactions."
    GridWise Architecture Council[1]

Transactive energy is thus a core component of the notion of peer-to-peer energy trading. To be sure, exchanges were envisioned prior to the rise of DLT wherein electronic exchanges would be used to facilitate such transactions. However, there is a challenge that you need to get the parties to a transaction to agree on a platform for such exchanges. Additionally, there are often regulatory restrictions that have limited the use of peer-to-peer energy trading that have only been lifted recently (within the last few years). In England, peer-to-peer energy trading has been available for some time – but this is a paper-based system and it was limited to large energy providers. Other pilots have been allowed, perhaps most notably in New York City, and in Australia and New Zealand, albeit other pilots have started to be allowed in other places around the globe. In Arizona, the public utility commission opened a docket so that people could provide information to the regulatory commission about DLT and its application in peer-to-peer energy trading. Regulators are trying to arm themselves with information as DLT disrupts how buying and selling occurs, not just of energy, but as we have seen, all kinds of goods and services. Commissions have been going slow – and no doubt are hesitant to allow a new technology to disrupt the status quo when there may be unforeseen consequences. The state of Texas is often held up as the example of successful deregulation in the electric utility industry, but California's energy crisis of 2002 is often held up as the counter example and was considered a failure of regulation.[2] "This concept sounds amazing" argument gets tempered by the "what if the wheels fall off" argument.

Back to the common platform question, consider the figure below and all of the players and boundaries that might exist in a transactive energy environment. Each one of these players and each one of the exchanges across boundaries would traditionally involve different applications or platforms to enable the exchanges. With DLT and smart contracts, a solution could be created whereby the assets are each registered so that their owner and services provided are documented, but also, potentially, a single platform could be used to facilitate the transactive energy exchanges between partners and between boundaries. A few companies already offer these types of services, such as Electron[3] and the Energy Web Foundation[4] (Fig. 14.1).

You may wonder how the architecture from the Transactive Energy Framework differs from the federated architecture discussed earlier. The short answer is: not much. The top right corner fits the federated architecture that was discussed earlier quite well. The same challenges exist.

---

[1] Transactive Energy https://www.gridwiseac.org/about/transactive_energy.aspx.

[2] The California Electricity Crisis: Lessons for the Future, James L. Sweeney, https://web.stanford.edu/~jsweeney/paper/Lessons%20for%20the%20Future.pdf.

[3] https://www.electron.org.uk/.

[4] https://www.energyweb.org/.

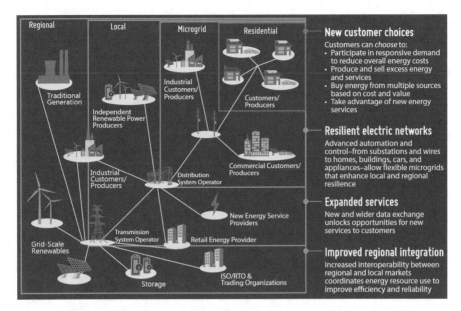

Fig. 14.1 Gridwise architecture council, transactive energy framework, role and boundaries

You need devices that have the computational capability to participate in the DLT ecosystem, but also require the necessary communications. This will most likely occur in denser population areas where there is more broadband communications available, before these capabilities make their way into the countryside with limited communications and fewer customers that will not make for an attractive market.

However, as we progress further down and to the left in the info graphic, you see larger, regional players. These sites typically already have computer systems capable of running or interacting with a DLT ecosystem, or have the capital available to invest in such systems, and already have significant communications capabilities available that already serve their legacy systems. Thus, these roles and organizations are more readily able to adapt to a DLT ecosystem. However, when you consider that these roles and organizations already have systems in place for exchanging data with their peers, a DLT system needs to provide additional value to justifying replacing the existing ecosystem. This takes us back to the "better, faster, or cheaper" conversation. These organizations are not going to change "just cause". There has to be a business driver and a clear ROI – and insight into what risks there may be, before the large organizations that are involved will be ready to make a change into how they operate unless, of course, directed by the regulatory community.

Energy Markets/Price Complexity

This brings us to the types of markets that exist in the energy sector. These markets are quite different than traditional retail operations where one sells a "thing", such as a phone, a car, or a hat. This is because that outside of some storage capability such as batteries or pumped storage facilities, as was mentioned earlier, electricity is consumed when it is created. You can try to forecast demand just like you would if you were a hat maker, but instead of stacking your hats in some closet, electric system operators need to match generation and load on a real-time basis. If they do not, bad things happen. Equipment can be damaged; power lines can trip off leaving customers in the dark.

Proponents of transactive energy believe that market mechanisms can be used to solve the supply and demand needs of energy throughout the day via a combination of long-term and spot market agreements. The long-term agreements might relate to capacity – making sure that there are entities available to call on should the need arise for additional electricity generation. The "spot markets" handle more short-term needs. These would be intra-day trades – one example is demand response-type solutions (demand response is where you promise to reduce your energy consumption a few times a year in exchange for more attractive electricity rates). Another example is referred to in the industry as "ancillary services". These types of services usually require the participant to be able to respond in seconds. Some ways that an entity might respond is to reduce the energy they are using (such as the demand response case), or curtail the energy that is being generated in a situation where we have too many electrons on the grid. Another example of an ancillary service is changing the energy characteristics being created – from watts to what is called reactive power, which can facilitate increasing "hosting capacity" which is how much energy a given electric circuit can support before something goes awry.

Distributed Energy Resources

Related both to the IIoT concepts and the federated architecture discussed previously, Distributed Energy Resources (DER) are any type of energy source that is not base load (think the large nuclear, coal, gas, hydroelectric, etc.) generation. DER is typically smaller (albeit we now see "utility scale" installations of renewable energy sources), but typically are solar, battery, and wind systems, although technically, small-scale micro turbines, or diesel generators can also be thought of as DER.

When a solar panel is put on the roof of your house, there is a device that converts that energy into something that can be used on the grid. The solar panel creates direct current (DC) energy, and this needs to be converted to alternating current (AC) energy and the frequency also needs to be matched to the grid, 60 Hz for North America, and 50 Hz for Europe and part of Asia. This device is called an "inverter". The first-generation inverters were not that smart. They did their conversion job and not much more. They had some capability to do some sensing of the local energy conditions and to react to them, for example, disconnecting from the grid if the power goes out. (You want the local inverter to disconnect from the grid to avoid electrocuting line workers that will be coming to make repairs.)

As time went on, it was recognized that these inverters needed some additional capabilities in order to deploy more renewable, distributed energy in the distribution grid. At high levels of deployment, DER can be disruptive, creating energy situations that are out of power quality tolerances that can damage equipment. But if communications could be added to the inverters, hence "smart inverters", more controllable capabilities could also be added to the inverters. A smart inverter can be told to limit how much energy a DER creates; it can be remotely disconnected (removing its connection from the grid) or it can be told to generate reactive power (again, which helps with over-generation – essentially when too many electrons are being produced), just as a few examples.

This is where the DERMS comes in. A DERMS is a DER management system.[5] Because as millions of these DER and their corresponding smart inverters are deployed, it is impractical to manage these devices one at a time, with the grid operators sending all those commands one at a time. With more and more devices, the job just becomes overwhelming. What is needed is for these to be managed in aggregate (groups) so that an operator can send out a single command and control hundreds, if not thousands of devices. And, because these smart inverters do not all speak the same "language" (due to competing or lack of standards), the DERMS needs to do all of the translation.

As originally created, a DERMS would be a system at the utility, much like the system that collects meter readings from meters. But it is also envisioned that there could be a "DERMS in a box" that could be deployed to neighborhoods or microgrids (see figure below).

In this figure, we see three neighborhoods. Potentially, each neighborhood could have a DERMS controlling the local DER. But more applicable to our conversation on DLT, this DERMS might also function as a node in a DLT network, with the individual smart inverters running wallets (Fig. 14.2).

These are cleverly shown with the wallet icons that send transactions to the DERMS, and the DERMS sending these transactions to its peer DERMS in the DLT network. This is much like the relationship we see in the federated model presented previously in Fig. 10.3. The Distributed DERMS (or dDERMS) is a computer that has the horsepower to handle the consensus algorithms of the DLT used to manage the transactions, as well as handle any command and control functions the individual DER may need to operate within the limits of the local energy network.

In this configuration, this matches the conceptual federated architecture that was discussed earlier. This also matches the notion that less computational capable

---

[5] From Research to Action | Advancing the Integrated Grid: Distributed Energy Resource Management Systems (DERMS), https://electricenergyonline.com/energy/magazine/866/article/From-Research-to-Action-Advancing-the-Integrated-Grid-Distributed-Energy-Resource--Management-Systems-DERMS-.htm.

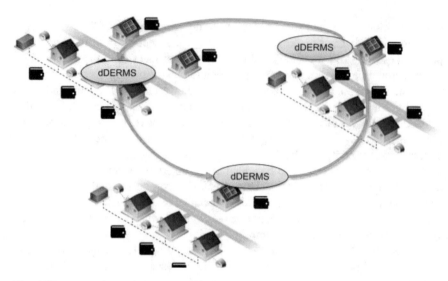

Fig. 14.2 Distributed DERMS concept with peer DERMS running the consensus algorithm for the DLT

devices only run enough software, such as a wallet, to interact with nodes; in this case, a distributed DERMS could also be running a node of a DLT. These nodes process the command and control necessary for the operation of DER, but could also serve as nodes in a transactive energy framework that was also discussed earlier. This sort of network enables the ability for prosumers that generate more electricity than they need, to sell it to other prosumers in the network that may have a need, with all of the exchanges tracked in the immutable ledger. The utility running the distributed DERMS could still step in with command and control to manage the DER, for example, in an over-generation situation where the DER might need to be curtailed.

Shortcomings: The creation of distributed DERMS is only now being developed and the DERMS market is, as of now, still maturing as vendors wrestle with implementing the necessary functions for grid management. Further, although there have been pilots that use special meters in conjunction with the utility meter for the purposes of interacting with a ledger, the cost of the meters that are capable of these capabilities raise the question of whether the benefits of the transactions would pay for the cost of the specialized meter over its lifetime (Table 14.1).

Table 14.1 DLT fit: Distributed DERMS

Characteristic	
Security	DLT would provide for greater security mechanisms, which could be layered upon existing smart meter communications
Immutability	DLT would provide an unchangeable record of transactions and audit trail for generated energy that could be used to verify the amounts of generated that are recorded with the RECs.
Transparency	Anyone could validate the transactions that occurred in the DERMS to verify energy measurement, command and control actions, and device configuration changes.
Relative to status quo	Distributed DERMS is only now being developed. Although leveraging some existing technologies for measurement (smart meters), this has not been developed yet in conjunction with DLT. *Opportunity:* Greenfield; this is a new area for development. If a vendor can solve parts of these problems, e.g., DER asset registration that is often a paper-based or otherwise unwieldy process – utilities should get on board. The same old rules apply, show value – make life easier for utilities and the money should follow. *Challenge:* The need to develop proof-of-concept that would leverage DLT, DERMS, and transactive energy capabilities, plus get these new capabilities sold in such a way that one can gain the support of the regulatory community.

Questions

1. DERMS refers to what?
2. A "smart" inverter is a type of IIoT device (True/False)
3. What is a "DERMS in a box"?
4. What is a prosumer?
5. What is a REC?

Chapter 15
Auditing: What DLT Was Made for

Learning Objectives
- Consider some ways that DLT and its immutable ledger could be used to complement auditing, safety, or other configuration management tasks

One use of DLT that often seems to be overlooked is its application for auditing. The immutable ledger seems to be tailor-made for such an application. What constitutes an audit? Anywhere there needs to be a check on activity – this could be the audit of financial records, or checklists, or it could even be things like tracking the configuration management of computers or other devices and assets. For the financial flavor of audits, consider what happened in the United States, which led to the passage of the Sarbannes-Oxley Act.[1] Due to shenanigans by companies such as Enron, Tyco International, Adelphia, Peregrine Systems, or WorldCom, steps were taken to require more robust reporting of financial systems in the hopes of preventing similar abuses. One could argue about the effectiveness of the implementation of the law. It certainly led to a lot of spending by companies to bring their computer systems and processes (either by upgrade or acquisitions) in line with the requirements of the law – and it certainly has made good money for the companies that are contracted to audit the transactions to ensure that companies are in compliance.

Companies in the United States spend millions of dollars with auditing firms for them to certify their records that they are accurate and that they are not in error or have been manipulated. If only there was some sort of technology that could be used to address this sort of challenge. For example, imagine if you had a record that could not be changed, that had a smart contract that recorded the transaction and the approver of the transaction, whether it is financial or a software change to a financial system. Such a thing seems like it might provide value to someone someday.

Change control associated with financial systems requires that when a change is made the person that makes the change cannot be the same person that approves the

---

[1] H.R.3763 - Sarbanes-Oxley Act of 2002 https://www.congress.gov/bill/107th-congress/house-bill/3763.

© Springer Nature Switzerland AG 2021
G. R. Gray, *Blockchain Technology for Managers*,
https://doi.org/10.1007/978-3-030-85716-5_15

change. This is to prevent someone from putting in unauthorized changes that could lead to fraud or other abuses. Change control systems these days are often automated to account for this separation of changer and approver via role-based security. A computer system is used to keep track of the identity of the people working on the system, and also their role. An example of such a system is Microsoft's Active Directory. Hopefully, the administrators of such a system are not giving themselves both "changer" and "approver" roles – sidestepping the controls that have been put in place. This is one of the functions of a financial audit – to ensure that the changes are tracked and that no one has done something to circumvent these processes and procedures.

Control Objectives for Information Technologies (COBIT), supported by the ISACA organization, outlines the auditing requirements and has a vast array of materials for various systems to support these auditing activities.[2]

These are some great tools and well-understood audit requirements. What does DLT bring to the table?

The immutable ledger.

These automated systems have control logs and access constraints, but chances are somebody, somewhere, knows how to circumvent these controls, or work with others to circumvent the controls. If change control logs were part of a smart contract/dApp as we have seen in earlier examples, there would be no question about the contents of such logs. The DLT ledger is immutable and any party that wanted to cover their tracks could not do it even if they wanted to.

Safety Checklists

Another kind of record that is related to audits is the checklist. Checklists are used for many types of work to help ensure that proper procedures are followed. These are usually developed to protect personnel and equipment. Maybe you have seen the pilots on your last flight going over their checklists as they ready the plane for take-off. Another example is when crews are working on electrical or other heavy equipment. If you have ever seen one, how often have you noticed that these are still paper and that someone has a binder open and they are reading through the list. Probably too often. Expert systems are changing this situation with voice-activated controls or "experts-in-your-ear", that is, if a person happens to be using a head-mounted computer, with a camera a remote subject expert can see what the operator sees. But again, if the checklist itself is paper, without controls, there will continue to be opportunities where people miss steps. And if a step is missed and equipment is damaged or people are killed or injured, an investigation ensues to make sure those procedures and checklists were adhered to.

---

[2] ISACA - AUDIT PROGRAMS AND TOOLS, https://www.isaca.org/resources/insights-and-expertise/audit-programs-and-tools.

Fig. 15.1 Immutable ledgers will permanently store audit trails

There may be an opportunity for a smart contract/dApp that could verify both the steps in a checklist are followed, and that associated equipment settings are tracked as they are made. If these changes are tracked in the immutable ledger of DLT, if a step is skipped that results in injury, death, or damage, there would be no editing of the log to cover such an occurrence. Such a system might need to be tied into whatever control system that associated equipment is using, or photographic or video evidence could be captured by head-mounted computers or smart devices (phones and tablets). Thus, as each step is executed, a permanent record is stored to ensure that procedures were followed (Fig. 15.1).

Configuration Management

Computers and other electronic devices, or assets that support communication have configuration data associated with them. Information such as the versions of the software they are running, the settings for the communication components, serial numbers of components, or the versions, sizes, and timestamps associated with the files on these devices are just a few examples of the configuration data that is stored. Configuration data is important for tracking changes to settings or components of a device. As the numbers of devices a company manages increases, this situation only becomes more important. If you only had a single device and something went awry, you might return it to have it repaired, or return it to the manufacturer. If you have to manage thousands or even millions of devices and you need to maintain the security and communications capabilities,your organization might be the one on the hook for maintaining the devices and handling unplanned for configuration changes.

Assume something has gone awry with a type of device in your inventory that you manage. You might wonder if it is something wrong with the hardware, or if something has changed with a configuration setting. Some file has gotten out of whack, or a change was pushed through that had an unintended consequence, or perhaps you need to push out a security patch to deal with a newly discovered flaw. Or, you need to check to see if there has been an unauthorized change to a devices configuration. There are many reasons to need to be able to track and compare configuration data.

DLT's characteristics seem to be a good match for addressing this use case. First, the security component could be used to ensure that changes were only allowed by those entities with the security keys that were authorized, and the communications for such changes were secured as well. This is a feature that traditional configuration management systems also provide – but with DLT and using a ledger, you know that this security is built-in. Traditional configuration management stores configuration data in a database to keep track of changes and the authorization for those changes. If something goes awry, some examples of what such a system provides are that operators can compare a configuration baseline to what a device is supposed to have, to ensure that it matches and that no unauthorized changes have occurred. The configuration settings of like devices can also be compared. They could also verify who or what made the last set of changes and when they occurred to track down if a change coincides with errors seen on a system or with the managed devices. What DLT gives you is another check in the system where device configuration changes are made; these changes are stored in the immutable ledger – there is no backdoor to go around the tracking of these kinds of changes (Table 15.1).

Table 15.1 DLT fit: Auditing

Characteristic	
Security	The security features can be used to ensure that only authorized persons or entities can make changes to a system – and that their roles correspond to the task that they are performing.
Immutability	The immutable ledger can be used to verify configuration changes, role changes, or steps that have been taken in a checklist.
Transparency	With an open ledger, any transaction could be reviewed and used to verify the change and the approver, whether that be a human actor, or a system.
Relative to status quo	*Opportunity:* For checklists which are often paper-based, leveraging a dApp could greatly improve speed, accuracy, and accountability – but would need to ensure that the investment into such a system did not outweigh the value of the people or systems being protected. *Challenge:* Some systems such as auditing and change management systems at large organizations subject to legislation such as Sarbanes-Oxley Act have existing systems – the value proposition of leveraging an immutable ledger would have to provide an ROI or risk improvement over the status quo. For configuration management systems, this is another situation where an organization may feel that their current system is good enough and the additional security or immutable record characteristics may be seen as overkill.

Questions

1. What was one of the drivers for the passage of the Sarbanes-Oxley Act?
2. The immutable ledger provides what sort of benefit for audits, configuration management, and checklists?
3. Change control requires that the person making a change cannot be the person _____ the change
4. If you had a mature and automated (approved changes were automatically deployed) configuration management system, would you change to one based on DLT?
5. Checklists are designed to protect _____ and _____

Chapter 16
B2B, Brokers, and Gigs

Learning Objectives
- Understand some of the opportunities and challenges that DLT presents for other types of B2B interactions

The privatization of the Internet led to a rise in e-commerce. Anyone that has shopped at Amazon knows that is not a surprise. But e-commerce and business-to-business (B2B) integrations of all stripes were also created. In this chapter, we will look at the example of an appliance service plan. But you could substitute any situation where a vendor does not perform the work contracted for themselves, but is in reality, a broker – they connect customers with the people that have the skills and equipment necessary to perform a service.

In the Appliance Service Plan use case, a customer worried that one or more of their appliances might fail, signs a contract with a provider that essentially offers insurance for one or more of their appliances. This is often bundled with a service to inspect certain assets such as furnaces and water heaters, with a guarantee that if an appliance under this extended warranty should fail, the replacement of the appliance is covered by the plan. This situation sounds like it has all the makings of a smart contract. And in fact, there may be smart contract systems in place that currently automatically executes these types of agreements when the terms are met – just not based on DLT. Before DLT, such systems might have been offered by your local utility, big box store that sold appliances, or other appliance retailers. With the rise of the Internet and e-commerce, such systems as these were developed and hosted on the respective web sites (and you could probably find some today).

Often with the Appliance Service Plan model the inspection and replacement work is not performed by the company with whom the customer contracted with, but is farmed out to other contractors. The contracting company merely serves as an agent – connecting the customer to the people that perform the service. In this case, there may be multiple contracts (all of which could be "smartened up") at play. There might be one contract for the person that is performing the appliance inspections, and another for an installer should one of the appliances needs to be replaced.

© Springer Nature Switzerland AG 2021 149
G. R. Gray, *Blockchain Technology for Managers*,
https://doi.org/10.1007/978-3-030-85716-5_16

This is great for the customer ("Yay – I do not have to worry about my appliances"), great for the broker ("Yay, I get a cut of every transaction and I can use sub-contractors to manage my work demand, and great for the contractors ("Yay – I do not need to beat the bushes for new work – work comes to me!"). You could even go one further and have the scheduling of any service work handled by the broker as well. This way the broker can make sure that the service contracts are well coordinated and meet the needs of the customer. These relationships are shown in the figure below (Fig. 16.1).

Now, could these all be smart contracts be instantiated in some form of DLT? Why, yes, of course. The benefit of a smart contract is that once the terms are met, the contract is automatically executed. Do these relationships need to be based on DLT? No. As previously mentioned, many forms of brokerage have existed pre-DLT and continue on today without the need for DLT.

What does DLT get you then? Maybe a contract in an immutable ledger is useful if a user is concerned that terms of an agreement might change. Have you ever received one of those messages that state something akin to, "The terms of our agreement have changed. If you don't agree to the change, stop using the service"? This sort of situation always seems a bit one-sided, yes? If the contract is in an immutable ledger, the contract is the contract. This sort of requirement is more for providers that cannot be trusted – but then, if you cannot trust them, why are you doing business with them? The other feature is security. Sure, your pocket super-computer (smartphone) is pretty secure, as smart devices go – with facial recognition or other biometric protections and/or locking features. But there would be the added security of the messages and identities that DLT would provide. Do you need

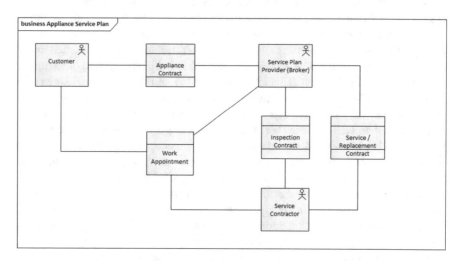

Fig. 16.1 Appliance service plan contracts and relationships

that to make sure your pizza is delivered? (This is another sort of brokered gig job as seen with new start-ups like Grubhub[1] and Doordash[2]). Probably not.

Other Gigs

The "gig" in gig economy borrows its name from the world of musicians. Usually, musicians are not employed by the places in which they play. These are short-term deals, usually only for a couple hours a night or so. Even in the case when a musician or a band has a recurring appearance with an establishment, they still are not employees of that establishment. They perform for the allotted time and get paid at the conclusion of their allotted time. In the last few years, we have seen a rise in more gig-type services. Uber[3] and Lyft[4] are the dominant players that provide ride-sharing services – where the driver takes on the gig of giving someone a ride to or from the airport, supermarket, or other destination. Other broker-type services have emerged for temporary stays in lieu of getting a room at the hotel, food delivery (as previously mentioned), and even dog walking[5]!

These app-based services are displacing or competing with legacy taxis, hotels, the pizza delivery person provided by the pizza shop, and... well I do not know what the dog-walking service displaces... your neighbor (probably much to their relief). The competition has mostly been a good thing. With the rise of Uber and Lyft, when I still took a taxi, I noticed a transition to them being in better repair... and they certainly smelled nicer. With increased competition from providers like AirBnB,[6] hotels have to provide more competitive pricing. The pizzeria does not need to hire a delivery person and can focus on making pizza. Pets get out instead of being holed up in their domicile wondering if their owners will ever return.

The news is not all good – in metropolitan areas such as New York City there are complaints that ride-sharing services have actually resulted in greater numbers of vehicles causing even greater congestion and air pollution. Complaints in the hotel industry range from unlicensed usage to noise complaints arising from increased use of properties or usage that gets more exuberant than normally allowed. Some of these complaints are legitimate, some are the type of noise one hears when legacy business models are displaced and the incumbents try to do everything in their power to dissuade the newcomers.

These services are all new and convenient, but is there room for DLT in the broker/gig model?

---

[1] Grubhub https://www.grubhub.com/.

[2] Doordash https://www.doordash.com/.

[3] Uber.com Get in the driver's seat and get paid.

[4] Lyft.com Hop in. Crack a window. Let's get back out there.

[5] Wagwalking.com Local, trusted pet care Book 5-star Pet Caregivers near you.

[6] https://www.airbnb.com/.

Table 16.1 DLT fit: B2B, brokers, and gigs

Characteristic	
Security	DLT could provide a secure means of protecting identity, privacy, and transmission of information associated with a broker or gig-type smart contract
Immutability	With the immutable log, all participants to a transaction should be able to see and verify the details of a said transaction and know that it has not been tampered with – although I think the risk of someone adding pineapples to my pizza without my permission is fairly low. (Although yes, pineapple does go on pizza, with ham and jalapenos).
Transparency	With an open ledger, any transaction could be reviewed and used to verify the change and the approver. This might be useful if there was any question about the terms of a contract.
Relative to status quo	*Opportunity:* Automating a smart contract with DLT – and the smart contract could be any type of brokerage or gig-based transaction. Is the real opportunity in providing a platform for *all* brokerage-type business models? *Challenge:* Probably overkill compared with how gig apps have been created. How much security and transparency does one need? Is there room for another implementation model? For the incumbent gig services providers, maybe not. For new entries, there may be an opportunity here.

Potentially. As we have discussed, enhanced security offered by DLT might be attractive to customers, especially any that might not want to give up privacy for convenience of not having to pick up that pizza themselves. Having an immutable ledger for automating orders and deliveries might also be attractive. Better yet, if one could provide a DLT platform that provides these broker services for anyone that has a service to offer that needs to connect with customers, thereby eliminating the step where every time someone comes up with a new gig idea they need to recreate all the services for account creation, security, privacy, order fulfillment, and delivery. As the old saying goes, "in a gold rush the only people that are sure to make money are the people selling the shovels", perhaps the company that provides the DLT brokerage platform will be the shovel sellers of the gig economy, much as those that sold web servers were the shovel sellers when the Internet revolution exploded (Table 16.1).

Questions

1. Connecting customers with service providers is referred to as a _____?
2. The company you contract for a service farms the service out to a third party, you are _____?
3. Who gets rich in a gold rush? Who gets rich in the gig economy?
4. Extended warranty service on your vehicle is one way to implement a ____ service?
5. Pineapple on pizza?

Chapter 17
Utility Metering

Learning Objectives
- The component parts that make up a typical utility metering system and where DLT might fit in such an ecosystem

Chances are, you have got at least one meter on the side of your house or apartment. Electricity, gas, water, steam – meters of all kinds, and they are used to measure the consumption of those commodities. More and more, older electromechanical meters have been replaced with so-called smart meters. (About 70% of households now have these smart meters[1]). Meters initially began changing from electromechanical devices to digital meters. As meters continued to advance, communications capabilities were added. The first round of smart meters were referred to as "automated meter reading" (AMR), because the initial goal was simply to eliminate the requirement that meters be read by a human. Later, additional capabilities were added to these meters that facilitate more two-way communications features to what is now referred to as advanced metering infrastructure (AMI) (Fig. 17.1). Some of these commands might be to connect or disconnect meters due to nonpayment/repayment, or in the process of transitioning to new home owners, reconnect without requiring a trip to a utility office to get an account set up. Electric meters, because there is a ready source of energy to run the metering and communications, can be run in an "always on" mode, whereas meters for other commodities use a battery to power the communications module. Thus, battery-powered meters only "wake up" to send meter readings, and then the communication module goes back to sleep until the next scheduled time occurs for readings to be sent. The different features of AMR and AMI meters are shown in the next figure.

In addition to the meters, for a smart metering system, there are several components involved in the processing of metering data. In addition to the meter, there is

---

[1] Smart Meter Deployment Projected to Reach 107 Million As of Year-End 2020, https://www.tdworld.com/grid-innovations/smart-grid/article/21120206/smart-meter-deployment-projected-to-reach-107-million-as-of-yearend-2020.

© Springer Nature Switzerland AG 2021
G. R. Gray, *Blockchain Technology for Managers*,
https://doi.org/10.1007/978-3-030-85716-5_17

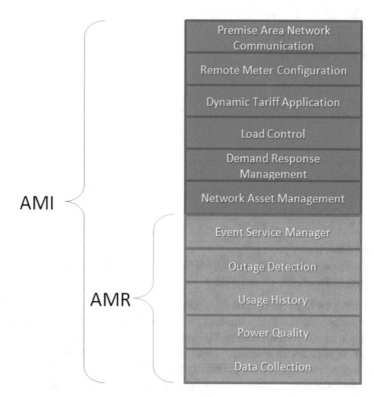

Fig. 17.1 The difference between AMI and AMR meters

usually a regional data collector. The collectors (by their name) gather data from the meters that are closest to them. They then send the data on to what is called an AMI "Head-End" system which aggregates all of the data from all the collectors. The utility may also have a meter data management system (MDMS) where other data analytics might be run; such a process is called validating, estimating, and editing (VEE). VEE checks to make sure that there are not any bad meter readings, and if there are, it makes an attempt to correct those errors via an algorithm. Sometimes, the VEE process is run in the customer billing system, other times VEE is run in the MDMS and then the data makes its way to the customer billing system. As you can see, there are many steps to get the data from the meter and then ensure it is of sufficient quality that it can be used to generate a bill for the consumption.

Utility metering systems, occasionally, are suggested for systems that might fit to be replaced with DLT, but there are a few challenges with in making that a reality. For example, meters do not have a lot of computational power as we discussed in the section about IIoT and inflated inspections. This is a challenge if one is going to use them as peer in the distributed network. DLT networks are very "chatty" (they talk a lot) and have storage requirements that will normally exceed the capability of your average smart meter. Potentially, while not being a validating node in the network, with the right configuration they might be configured to run a wallet to transact on the network – each of those meter readings becomes a transaction, and they could be bundled up into blocks by the collectors. (Although collectors tend to just be "store and forward" devices, themselves not with much computing power). If you squint at such a system, you might think, "oh yeah, that could work".

There is another challenge in using a smart meter as a peer or device in a DLT network. Smart meters typically have a lifespan of 12–15 years. Consider this mental exercise: walk into your local big box store and consider the computers that they have for sale. Now, all you need to do is pick one that you know will run reliably for the next 12–15 years, be secure, be able to support remote updates (for security patches in the future or new functionality), and be able to run continuously without being rebooted or being taken into service. Choose wisely! To be fair, your average laptop computer is way more complicated than a smart meter – the more functions something can do, the greater the potential for failure. Additionally, one of the value propositions for a meter is that the price be kept as low as possible. If personnel have to be sent even one time to check on a meter, the meter just got a lot more expensive in terms of total cost of ownership. Thus, you cannot be sending technicians out to reboot this computer – it has to be reliable.

There are also other communications challenges when you consider the network capabilities in urban areas versus rural areas. In the city there is connectivity everywhere – although maybe it is not perfect due to interference from buildings and foliage, but it is everywhere. In the country, sometimes it is not even there. Or if a communications capability exists, it does not support the bandwidth required for the communications traffic that DLT would create.

There is one benefit for the utility if one could solve all of the technical hurdles: plausible deniability (Fig. 17.2).

Consider the "high bill complaint" use case. A customer calls in, irate that their bill is so much higher than previous billing cycles. The dutiful and helpful customer service representative (CSR) will "ping" the meter (i.e., they will remotely connect to the meter and pull the latest data from its register) to check communications and verify the reading.

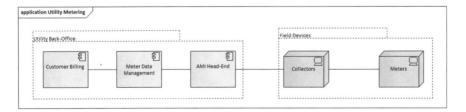

Fig. 17.2 Data flow and systems involved in utility metering

And then what happens? The CSR verifies the readings, and the customer still has a high bill. But wait, the customer calls the utility, the utility checks *their* data system, and assures you that, "no sir or ma'am, the computer didn't make any mistakes". If the metering data was moved into a DLT ledger, with numerous validators, well it is not the *utilities* system now any more is it?

However, if the data is in an immutable ledger, the utility could not change the data even if they wanted to (they might create a new smart contract to give the customer credit for a variety of reasons – but the original record would not be changed). Are utilities jacking up customer's bills just to get a few more dollars? No. That would never survive an audit. But mistakes happen. And sometimes data that comes from a meter gets screwed up. Remember the VEE process that attempts to correct for missing or bad readings? You might challenge the algorithm and how it "fixed the glitch". And it is not unheard of for a VEE process to get fooled. For example, some utilities run a demand response program. For some reward to the customer, the utility can send a command to your home systems that will decrease how much energy is being used. It might be as simple as turning off a pool pump or perhaps changing the thermostat setting by a couple of degrees. What happened was that when the VEE process saw this dramatic drop-off in consumption, the algorithm "fixed the glitch" and averaged out what it thought the load was supposed to be. Whoopsies! Of course, that situation was corrected, but this example just goes to show that no system is perfect, usually because those pesky humans that run it cannot think of everything.

One benefit from DLT communications is the ability to digitally sign a transaction. Or to also only allow communications from components in the system that match the digital identity that might be used with DLT. Except that metering systems usually employ similar, if not the same security safeguards. They may not be part of a peer-to-peer DLT network, but chances are that they use similar public key-private key encryption capabilities. Thus, we run back into the argument of "How much better would the security be and at what cost?" Chances are given today's capabilities; the payback of such a change would be difficult to reach, even given a meter that had a 12–15 year life span.

At the end of the day, customers need to trust that the utilities metering systems are accurate and that their bills reflect actual usage. That trust might be achieved in better ways than switching out the metering system to one based on DLT (Table 17.1).

Table 17.1 DLT fit: Metering

Characteristic	
Security	DLT would provide for greater security mechanisms for metering communications. However, the question would be for legacy metering communications, if the current security is "good enough". How much better and at what cost?
Immutability	DLT would provide an unchangeable record of transactions. Physical device security would be required to ensure that what was written to the log is what was intended.
Transparency	The ledger would provide an open log of all transactions, command and control actions, measurement, and device configuration changes. However, electric meters currently can store changes in their internal register. The only difference being that nominally this register can be changed, whereas with a ledger, that record could not be changed (or deleted).
Relative to status quo	There would be a more difficult "better, faster, cheaper" story for legacy smart metering systems. They already have configuration and security controls built-in. A vendor offering a DLT solution in this space, where so far we have seen that the meters that can support DLT are more expensive than their non-DLT counterparts, makes this an unlikely avenue for DLT success. *Opportunity:* One value proposition may be in the realm of trust. Consider the use case of a high bill complaint. The customer calls the utility about a high bill. The utility investigates and may "ping" the meter to determine if it is working correctly. However, the customer may be unsatisfied with the investigation. After all, it is the utilities system that is being checked. But if the data used for billing was stored in a third-party ledger that may add an element of trust where the utility could not change the record (immutability), even if they wanted to. However, there would still be the problem of determining whether the smart meter wrote the *correct* data in the first place. *Challenge:* The computational capability of the device may limit what DLT capabilities it can use. Adding additional capabilities will drive up device cost.

Questions

1. What is the difference between an electromechanical meter and a smart meter?
2. Smart meters are installed in roughly what percentage of U.S. households?
3. How long is the usually life span of a meter?
4. What are some of the challenges with using a meter in a DLT network?
5. Can metering data be corrected once it has been written into an immutable ledger?

Chapter 18
Intellectual Property and Privacy

Learning Objectives
- Understand the characteristics that make DLT a beneficial platform for storing and tracking ownership
- Understand the difference between data owners and data stewards
- Revisit the notion of anonymous versus pseudo-anonymous

Intellectual property management and credential management are areas where DLT should shine. And DLT is not just suitable for intellectual or digital properties, but also for tracing ownership of actual property as well. In their article *Blockchain in Developing Countries,* the authors note that given that 42.9% of households in the developing world have Internet access,[1] and the 2011 U.N. report, *Corruption Leading to Unequal Access, Use and Distribution of Land*[2] that 61 countries had weak governance "had increased the likelihood of corruption in land occupancy and administration". The authors also noted that various reports had shown that it was not uncommon for officials to alter property titles or accept bribes for property titles. In theory, registering properties could be accomplished via a DLT-enabled app on a smart phone. People with such capabilities could potentially circumvent alterations to titles or other documents associated with property ownership. Such a system, with its immutable ledger, could certainly trace when and by whom properties were registered, and if written appropriately, would prevent the unauthorized transfer of property. Of course, such a system makes several assumptions, for example, that the people, if they could not use the system themselves (either a lack of smart device or the lack of understanding of how to use one), could trust the people that are working on their behalf. This goes back to the earlier discussion – if the

---

[1] ICT Facts and Figures 2017, www.itu.int/en/ITU-D/Statistics/Documents/facts/ICTFactsFigures2017.pdf.

[2] Corruption Leading to Unequal Access, Use and Distribution of Land, https://news.un.org/en/story/2011/12/397982-corruption-leading-unequal-access-use-and-distribution-land-un-report#.WEMpP33QCW1.

© Springer Nature Switzerland AG 2021
G. R. Gray, *Blockchain Technology for Managers,*
https://doi.org/10.1007/978-3-030-85716-5_18

system allowed for bad data (either bad faith actors [people], or compromised devices), the ledger will only be immutably recording bad data. And if there are corrupt actors, what is to stop them from using other means of coercion to force people to give up or otherwise transfer their rights – if prior laws were inadequate to prevent corruption. Still, the authors note that there are some trial projects, for example, in the countries of Ghana, Georgia, and India.

These examples were for "real" property. But DLT is ideally suited for intellectual property tracking. Think of all the different kinds of intellectual property, patents, copyright (books, movies, music, or other written works, etc.), trademarks... the list goes on and on. Each of these use categories of IP uses desperate mechanisms for tracking ownership. Some of these are still paper-based. Almost all of them involve "lawyering". Many forms of IP storage have moved to the Internet – but again, these systems are usually separated by category. And although one may trust that once the IP is stored in these systems, that the data would (hopefully) not be altered by unscrupulous actors. This is again the strength of DLT. Data is stored in an immutable ledger which anyone can review. In the case of IP, such a system would make it easier to determine ownership and any transaction that involved the transfer of ownership – all traceable in an open ledger. If an unscrupulous actor managed to make a change, when and by whom such a change was made would be available to all parties.

Privacy Management

Given all the security baked into DLT, managing privacy is another area where this technology can shine. We have seen some DLT such as Monero or Zcash, DLTs that are based on zero-knowledge proofs provide true anonymity. Recall that a zero-knowledge proof is a mechanism whereby a person can demonstrate that they know a secret without revealing what the secret is. In the case of DLT, the secret is your identity (why you are practically a super hero!) and the mechanism with which to secure it. Recall that cryptocurrencies such as Bitcoin are only pseudo-anonymous – and your identity may not be fully protected. Especially as we saw with exchange sites such as Coinify that required you to identify yourself in order to be able to use their exchange.

But you, dear manager, you may be wondering, if your business model requires the exploitation of customer's data, how you are supposed to make any money with everyone protecting their privacy? Privacy concerns have been gaining in attention – and have probably been more in the spotlight since the European Union passed the General Data Protection Regulation (GDPR) in 2016.[3] The primary goal of GDPR is to make it easier for users to protect their data – and also to create a

---

[3] Regulation (EU) 2016/679 of the European Parliament and of the Council, https://eur-lex.europa.eu/eli/reg/2016/679/oj.

regulatory framework for the EU that was consistent across the member countries on how data was to be treated – including when data was moved outside of the EU.

Probably the most visible impact of GDPR is related to cookies that are used to track users that visit web sites. A cookie is simple a small piece of data that is stored on your computer. They were originally designed to keep track of various states, for example, to keep track of where you were at during a browsing session on a web page. Then new kinds of cookies were developed (still small data files) that are used to track a variety of marketing data. The cookies that are used by the host web site that you visit are referred to as "first party cookies", whereas cookies that are embedded from other web sites and organizations are called "third party cookies". With GDPR, many web sites now announce the fact that they use cookies and give you an option to manage them. Of course, the default option is usually not "disregard all third party cookies"; the default option is to accept the use of all cookies. And to be sure, first party cookies can be important to helping ensure that your browsing session is consistent – but third party cookies are only for tracking your behavior. You could also use a browser; Brave[4] is one example, where privacy is the default, and it blocks cookies regardless of GDPR settings, although you can also manage privacy settings within the browser (Fig. 18.1).

Prior to the passage of GDPR, there was a similar attempt to standardize the authorization of the use of customer data by the customers, and a mechanism to provide the data in a standardized format. There are a few things to be learned from this effort. This is colloquially known as "Green Button". Green Button began life as the "Open Automated Data Exchange" after a White House based pushed in 2012, that users should have the ability to receive their meter usage data in a standardized format. This effort was ratified as a North American Standards Board (NAESB) standard, Energy Services Provider Interface (ESPI). The Green Button standard not only created a standard format for metering data that could be used by

Fig. 18.1 Privacy – easy to waive; harder to protect

---

[4] Brave browser, https://brave.com/.

customers (if you ever wanted to look at such a thing – likely not), but also provided a mechanism for customers to authorize third parties access to their data (something a third party might want to do). The premise is that third parties could potentially run analytics on the data and provide services based upon it. For example, determine how the customer might be able to reduce their energy bill. The important aspect for this privacy discussion is that in the grand debate about who "owns" the data it was recognized that although the utility gathered the data, the data was the customers – utilities were merely the stewards of that data, and as such, they had a responsibility to protect it and not disseminate it to third parties without authorization. The customers were the data owners. Green Button created a mechanism for the authorization of data sharing and also a mechanism to limit to whom the data could be shared and for how long the use of their data was authorized.

> Concepts like Blue Button and Green Button borrowed from the Staples (office supplies) "easy button" campaign. The message conveyed that accomplishing something should be as easy as pressing a button. The Staples campaign was very successful and led to copying by the metering and medical records initiatives. Everybody should have an easy button. This led one of my colleagues to quip, "my favorite button is cyan, because it was "C-Y-A" right in the name".

> Nothing enables C-Y-A like the immutability and auditability of DLT.

Consider other areas where privacy is a big concern – medical records, and their control and release. The Health Insurance Portability and Accountability Act (HIPAA) was passed in 1996 to provide requirements and rules around the sharing and confidentiality of medical records. The major goal of HIPPA is to "ensure that individuals' health information is properly protected while allowing the flow of health information needed to provide and promote high quality health care and to protect the public's health and well-being."

Additionally, medical records information has its own "button" (blue button[5]) with a goal of making medical records easily downloadable in a usable format.[6] Blue Button was originally created to support veterans receiving care under the Veterans Administration. For veterans that often changed duty stations, and hence, doctors and other care givers, maintaining a history of care was a bureaucratic nightmare. Blue Button, by specifying a standardized format and with Blue Button +, created a mechanism for sharing that medical information.

---

[5] Open Government Initiative, Blue Button, https://web.archive.org/web/20120203042534/http://www.whitehouse.gov/open/innovations/BlueButton.

[6] Blue Button Download My Data, https://www.va.gov/bluebutton/.

Consider another privacy-related restriction. The Right to Financial Privacy Act (RFPA) requires that the government must receive the consent of the customer before they can access said customer's financial information. Prior to this, the customer financial data, unlike the metering data example, was held to be owned by the financial institutions, not by the customers in a Supreme Court decision,[7] and that they were not subject to fourth amendment protections.

Another example of legislation that contains privacy stipulations is the Fair Credit Reporting Act (FCRA). This legislation was passed to promote the accuracy, fairness, and privacy of consumer information contained in the files of consumer-reporting agencies. Nominally, some of the provisions of this act are also supposed to help prevent identify theft, such as requiring the credit reporting agencies to provide customers with one free credit report every 12 months – this gives customers visibility to any shenanigans that might have been occurring without their knowledge.

In the utility industry, the New York public service commission adopted a "15/15"[8] rule for aggregated dataset use cases. This means that an aggregated data set may be shared only if it contains at least 15 customers, with no single customer representing more than 15 percent of the total load for the group. Additionally, a rule of "4/50" was created for building energy data aggregation standard. This rule requires that an aggregated data set may be shared only if it contains at least four customers, with no single customer representing more than 50 percent of the total load for the group.

One advantage for dApps that follows rules such as those instituted by the state of New York is that the process is less cumbersome for getting access to larger quantities of data. Although standards such as Green Button facilitate customer's ability to grant access to their user data, permission needs to be sought from willing customers. With properly anonymized data, with the appropriate control and approval (which can also be implemented in smart contract), the level of effort required to get access to amounts of data required to run analytics on is much reduced. If a vendor wishes to create services for a single customer's data, then the requirements of a standard like Green Button would still need to be met.

A benefit of having preexisting efforts such as Green Button, Blue Button, and other legislative sources is that the data requirements and rules are already codified in the respective standards and laws. If one was to make a platform for protecting privacy and authorizing third parties, they do not have to navigate legislation or work with customer groups to review ground that has already been covered. The rules and regulations are ready to be turned into code requirements by developers into any DLT platform that supports the creations of dApps. Sources such as these could be fertile ground for any DLT start-up that is looking to get into privacy protection.

---

[7] United States v. Miller, 425 U.S. 435 (1976), https://supreme.justia.com/cases/federal/us/425/435/.

[8] New York State 20-M-0082 Data Access Framework.

Table 18.1 DLT fit: privacy management

Characteristic	
Security	Leveraging the built-in security of DLT, users can be confident that their data that is stored on-chain is secured. As various hacks have demonstrated, it is the off-chain data that is the challenge, but this is essentially the status quo relative to other mechanisms
Immutability	Knowing that unauthorized third parties cannot tamper with the authorization mechanism should provide users with confidence that their data has been protected by the primary users.
Transparency	Having the ability to audit and confirm who was authorized when, not just be owners of the system, but by the data owners themselves will be an important feature.
Relative to status quo	*Opportunity:* The organization that makes privacy management easy wins. Protecting your privacy is always harder than waiving your rights. It always seems to take more "clicks" or it takes the use of additional tools, designed with privacy in mind "from the ground up", to protect oneself. But making privacy management user-friendly will be the key to success. *Challenge*: The main challenge with privacy management with DLT is the pushback from parties that do not want you to have effective privacy management – the primary culprit in this case is government actors, and the organizations that consume customer data that would prefer it not be regulated at all.

There is a potential privacy problem with DLT in the case where a participant in a transaction has willingly shared their identity. In some places, notably Europe, there are privacy laws that include the "right to be forgotten", that is, your data must be deleted. This poses a challenge if your transactional data has been immutably written to a ledger. Perhaps there might be a way to "pseudo-delete" the data via a smart contract – marking the customer data as irretrievable. This would perhaps honor the spirit of such laws, although it would not be complying with the letter of the law. Now in the case of completely anonymous DLTs, there would obviously not be a way for a third party to even determine if the customer data had been deleted since no one else would be able to determine which accounts were involved unless the person gave up the keys to their completely private and anonymous transactions. This would seem to defeat the purpose of having created the account in the first place (Table 18.1).

Questions

1. An identity that uses zero-knowledge proof is truly anonymous (True/False)
2. A cookie is _____
3. If the data being written to the immutable ledger is incorrect, the ledger fixes it before it is stored (True/False)
4. What was one of the benefits of the Right to Financial Privacy Act?
5. The 15/15 rule requires what sorts of restrictions on customer data before it can be shared?

Chapter 19
DLT Standards

As the old joke goes, "Standards are great. That's why everybody has one". One of the early criticisms of DLT, in addition to not scaling well, is that there was a lack of standards in the space. That is a fair criticism, but a situation is not to be unexpected when one understands the evolution of technology and the development of standards. Standards do not lead technology developments, they lag. Or, better described, they evolve from one level to the next. Consider this timeline, at Point A, a standard exists, then there is a "eureka!" moment (Point B) when someone develops an innovation. At Point C is the "piling on" period when, if a technology looks promising, similar solutions are developed. At Point D is the shaking out period. Vendors fold, are acquired, or otherwise exit the space. Eventually, the market evolves into a handful of (or in the age of the platform wars, usually only two) major players (Point E) (Fig. 19.1).

Usually, it is a Point D when there is also a rise of standards. Sometimes, in part driven by the need to establish a competitive advantage, especially when one is attempting to create an ecosystem based on "your" standard (as opposed to anyone else's), other times driven by the user community that wants to use the innovation with other things. This requires that the interfaces are well-defined so that additional integration can be developed. Once a standard is established, it is usually only a matter of time before the whole cycle repeats itself with a new innovation.

One example where this pattern of innovation and standardization cycle played out very quickly is what happened during the "browser wars" and the development of the hypertext markup language (HTML). In the early days of the World Wide Web, two browsers quickly dominated the market: Microsoft's Internet Explorer and Netscape Navigator[1] (Table 19.1).

During this time, new tags (the parts of HTML that tell the browser how to display information on the screen) were being added as innovations ahead of the respective editions of the HTML standard. Tables, colors, and frames, scripting, and

---

[1] https://www.w3.org/People/Raggett/book4/ch02.html.

© Springer Nature Switzerland AG 2021

G. R. Gray, *Blockchain Technology for Managers*,

https://doi.org/10.1007/978-3-030-85716-5_19

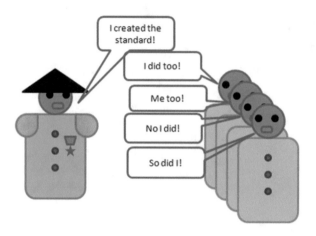

Fig. 19.1 Standards are great. Everyone has one

Table 19.1 "Browser wars"and the HTML standard

Year	Update
1994	HTML 2 aka "HTML +"; IETF sets up HTML task force; Netscape is formed
1995	HTML 3 published as a draft, tables added in v3.2; Internet Explorer comes out
1997	HTML 3.2 published
1998	HTML 4 published

applets are just a few examples. These innovations would be introduced by the browser makers, and then, discussions within the working group would eventually lead to these features being added to subsequent editions. This also highlights a new for standards to have an extensibility mechanism so that when new updates are added, they do not "break" implementations of prior versions. This is but one example of the standards/innovation cycle illustrated in the figure below.

The standards process is managed by various standards development organizations (SDO). You may be familiar with some such as the Institute of Electrical and Electronics Engineer (IEEE) or the International Organization for Standardization (ISO) or the Internet Engineering Task Force (IETF), upon whose RFCs (request for comment) are the standards that underpin the Internet. Some SDOs such as the preceding are international, while others may be regional, for example, the North American Energy Standards Board (NAESB).

There are other things that may not technically be standards, but are in such wide use that they are de facto standards. For example, although Google's G suite has captured the majority of the office suite market as of October 2020 at 50.6%,[2]

---

[2] https://www.statista.com/statistics/983299/worldwide-market-share-of-office-productivity-software/.

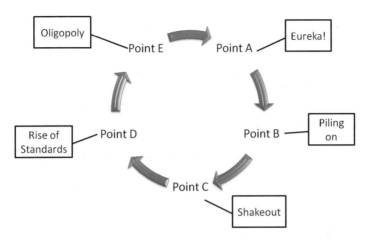

Fig. 19.2 Standards development cycle

Microsoft is still a dominant player and having been at the top of the market for so long, the file formats for Word documents (.docx) and Excel spreadsheets (.xlsx) are de facto standards, even though one can save Word and Excel files using one of the Open Document Standards (Fig. 19.2).

The discussion on standards is important, although we do not intend to be exhaustive in this chapter – there are a few important efforts underway. It is important for managers to understand what sorts of standards are in play in a given area. Vendors may make a standards compliance claim that managers need a mechanism to verify, or managers need to understand why a vendor may choose not to be compliant with a standard and what the ramifications of that lack of compliance might be.

A standard may not be published yet, as we will see with some of the efforts below. Thus, a vendor may not want to spend money chasing a thing that is likely to change, preferring to wait to see if a standards effort makes it to publication. It can also be expensive to have a product certified as being compliant with a standard and a vendor might want to spend those dollars elsewhere. A vendor may also not see the need to comply with a standard if their product has achieved some level of market dominance and is in fact a de facto standard.

Let us look at some of the standards efforts that are underway related to the development of DLT.

Institute of Electrical and Electronics Engineers

The Institute of Electrical and Electronics Engineers (IEEE) covers a wide range of standards that cover various facets of computing and electrical domains. Like other standards, the IEEE has a numbering convention for their standards. For the blockchain and DLT effort, it is IEEE 2418. From the IEEE:

This standard provides a common framework for blockchain usage, implementation, and interaction in Internet of Things (IoT) applications. The framework addresses scalability, security and privacy challenges with regard to blockchain in IoT. Blockchain tokens, smart contracts, transaction, asset, credentialed network, permissioned IoT blockchain, and permission-less IoT blockchain are included in the framework.

https://standards.ieee.org/project/2418_1.html

The original work effort was approved in 2017 via a PAR. A "PAR" in IEEE lingo is a "Project Authorization Request". Within the project various task forces may be organized into various related efforts. Such an organization is shown in the figure below that reflects related efforts for the effort around cyber security (Task Force 1), energy provenance, and certification (Task Force 2), Transactive Energy (Task Force 3), and distributed energy resources (DER) and electric vehicles (Task Force 5) (Fig. 19.3).

Related efforts are also denoted by the ".x" suffix as shown in the figure below. In this example, the IEEE standard 2418.5 is related to IEEE 2418.1, but is specific to the creation of an "open blockchain energy framework". The figure below shows how the IEEE 2418.5 standard effort is conceptually related to other efforts: IEEE 1547-2018 – *IEEE Standard for Interconnection and Interoperability of Distributed Energy Resources with Associated Electric Power Systems Interfaces*, IEEE 2030-2011 – *IEEE Guide for Smart Grid Interoperability of Energy Technology and Information Technology Operation with the Electric Power System (EPS), End-Use Applications, and Loads*, and P825 – *Guide for Interoperability of Transactive Energy Systems with Electric Power Infrastructure (Building the Enabling Network for Distributed Energy Resources)* (Fig. 19.4).

Within IEEE 2418, a reference architecture model has been created as shown in the figure below. It is a layered model that rests upon the devices that belong to the peer-to-peer network and the communications network. The "platform" layer is where various examples of DLT technologies are noted such as Ethereum, Bitcoin, and IOTA. At the higher levels of the model is where the services, processes, and data models one might be concerned with are shown. At the top layer is the distributed application (DApp). DApps use one of more services, processes, and data models to interact with the respective DLT platform. When considering using a DLT platform, the manager may want to have their technical teams dig into the details about how each of these is used; what languages are required, whether the data model itself is based on a standard (e.g., industries such as chemicals, banking, and electric utilities have their own common data models) and what services are

Fig. 19.3 IEEE 2418 Standards organization hierarchy

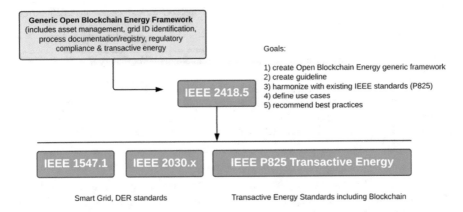

Fig. 19.4 IEEE 2418 standard in relation to other IEEE standards

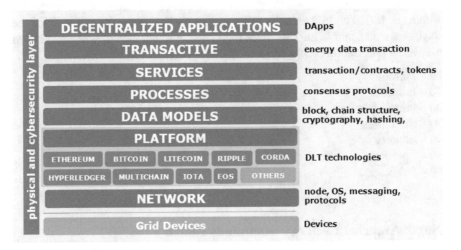

Fig. 19.5 IEEE 2418 layered DLT architecture model

exposed – whether developers need to code these services or whether the product that you are considering handles all of these details on your behalf (Fig. 19.5).

International Organization for Standardization

The International Organization for Standardization (ISO) technical committee 307 (TC307) has published three standards, with an additional 11 more being worked on as shown in the following tables.

Table 19.2 Published ISO DLT-related standards (https://www.iso.org/committee/6266604/x/catalogue/p/1/u/0/w/0/d/0)

Standard	Title
ISO 22739:2020	Blockchain and distributed ledger technologies – Vocabulary
ISO/TR 23244:2020	Blockchain and distributed ledger technologies – Privacy and personally identifiable information protection considerations
ISO/TR 23455:2019	Blockchain and distributed ledger technologies – Overview of and interactions between smart contracts in blockchain and distributed ledger technology systems

Note: The format of the standard number, like that of the International Electrotechnical Commission (IEC), has the standard number, with a suffix of ":YYYY" which indicates the year the standard was approved.

Each of these approved standards has been published only recently (Table 19.2).

The following table shows the further 11 standards that are under development within the ISO. The ISO uses the following numbering convention to indicate the stage that a standard is in. This gives the viewer a rough idea of when a standard may reach publication (Table 19.3).

00 – Preliminary
10 – Proposal
20 – Preparatory
30 – Committee
40 – Enquiry
50 – Approval
60 – Publication

North American Energy Standards Board

The North American Energy Standards Board (NAESB) is a non-profit organization that "serves as an industry forum for the development and promotion of standards which will lead to a seamless marketplace for wholesale and retail natural gas and electricity, as recognized by its customers, business community, participants, and regulatory entities."[3] NAESB announced the intent to hold a workshop to consider the development for a blockchain-related standard for the "development of a smart contract for the sale and purchase of natural gas as well as the development of supportive standardized modeling language that can be used in distributed ledger technologies."

The stated focus was to model the existing natural gas settlement and "post-trade" actions, then presumably, use smart contracts to automate facets of, if not the

---

[3] https://www.naesb.org//pdf4/092018press_release.pdf.

Table 19.3 ISO-related DLT standards under development (https://www.iso.org/committee/6266604/x/catalogue/p/0/u/1/w/0/d/0)

Standard	Title	Stage
ISO/CD TR 3242	Blockchain and distributed ledger technologies – Use cases	30.00
ISO/AWI TR 6039	Blockchain and distributed ledger technologies – Identifiers of subjects and objects for the design of blockchain systems	10.99
ISO/AWI TR 6277	Blockchain and distributed ledger technologies – Data flow model for blockchain and DLT use cases	10.99
ISO/WD TR 23249	Blockchain and distributed ledger technologies – Overview of existing DLT systems for identity management	20.60
ISO/DIS 23257	Blockchain and distributed ledger technologies – Reference architecture	40.20
ISO/DTS 23258	Blockchain and distributed ledger technologies – Taxonomy and Ontology	30.60
ISO/AWI TS 23259	Blockchain and distributed ledger technologies – Legally binding smart contracts	20.00
ISO/TR 23576	Blockchain and distributed ledger technologies – Security management of digital asset custodians	60.00
ISO/DTS 23635	Blockchain and distributed ledger technologies – Guidelines for governance	30.60
ISO/AWI TR 23642	Blockchain and distributed ledger technologies – Overview of smart contract security good practice and issues	20.00
ISO/WD TR 23644	Blockchain and distributed ledger technologies – Overview of trust anchors for DLT-based identity management (TADIM)	20.60

entire process. It turns out that according to the Department of Energy (DOE) the NAESB standard short-term natural gas trading contract is used for about ~90% of the transactions in the United States. It had been developed in the 1990s, and had a lot of legal "eyes" on it – so the contract language and associated data requirements had strong industry consensus backing it up. This was a case of where there was a natural fit and value proposition for creating a smart contract based on an existing contract.

The value proposition for moving the traditional contract to DLT-based contract is that the settlement process takes ~21 days. Considering the time value of money, automating the process using this technology makes a lot of sense.

To put this in place, NAESB created the framework for how the smart contract should be implemented such that it could be implemented in a DLT-neutral way, that is, it would not favor any DLT architecture or platform over any other. The details of this framework can be found in NAESB standard, *WGQ6.3.1 Contracts, Standards, and Models.*[4]

NAESB is working with the DOE to create an implementation test based on this framework to better understand the impacts to reconciliation and cyber security. To this end, beginning in 2021, a private, permissioned ledger will be set up, following

---

[4] https://www.naesb.org/wgq/cont.asp.

the standardized process and responding to cyber security threats, and tracking various metrics associated with the performance of the smart contract. NAESB will be the manager of the ledger.

If this implementation of natural gas short-term trading contract is successful, it will demonstrate how current real-world processes can leverage DLT in a way that has immediate benefits.

Questions

1. Innovation leads or lags standards?
2. What is the great thing about standards?
3. IETF is the _____ _____ _____ _____
4. The IEEE is the _____ _____ _____ _____
5. Products need to be compliant with a standard to be successful (T/F)?

Part IV
Questions for Managers

We have taken quite the journey through technological time have not we? We have learned that DLT did not spring from whole cloth, but has been built upon a whole host of technologies developed over the last few decades. Digital cash, digital identities, verified logs, proof-of-work, consensus algorithms... it is quite the daunting list. Are not you glad you had an easy mechanism for walking through each building block one topic (this book) at a time?

Now that we have walked through them, I should remind you, dear reader, that we only skimmed those topics. Different facets of these building blocks are often within their own realm such as cryptography or systems integration. They are not just "topics" but entire fields of study of themselves. But now you are armed with the fundamental understanding of the technologies underpinning DLT. With this fundamental knowledge, you can ask pointed questions of prospective vendors shopping their DLT wares, and also give you the basis to judge the different types of DLTs suitability for any given use case.

So, maybe you are not an expert – but you can definitely wow your friends at dinner parties! "Hey Dan, have I told you about the latest mechanisms of consensus algorithm used by this new DLT implementation?" Trust me – you will be a party favorite. All kidding aside, as interest has increased from the investment side of things, having a deeper understanding of why one DLT seems to be receiving more attention over another form would likely make you a legitimate center of attention as you attempt to explain these complexities – perhaps by using one of the use case examples that we explored in the earlier section.

Chapter 20
Matching DLT to Business Problems

Learning Objectives
- Review and reflection on the material; building blocks and use case applications and assessments

This is why we are here right? Now that we know a little something about what underpins DLT, we can evaluate the core characteristics: security, transparency, and immutability, and think about how these get matched to the business problems that are in front of us. You might even draw different conclusions based on your understanding or experience with how some of these capabilities have been implemented. That is actually OK. This mindset demonstrates that you have internalized your understanding of facets of the technology and how they might be applied. If so, the mission is accomplished from my point of view. In the previous section that explored various use cases, what came to mind about how you might apply the technology in your area? Was there something that struck you as directly applicable? Or was there something that might be applied in a related field, for example, a different application of a gig or brokerage offering?

Or perhaps this is your first time thinking through the technology and these example business problems. This is good too – the goal here is not to say, "use or do not use DLT with these specific cases", but rather, give you some examples of things to consider when you are evaluating your own business cases back at your organization.

Most likely you will not be developing your own DLT applications, but rather fielding sales pitches from prospective vendors that hopefully have figured out how to relieve some of your organizations' pain points. However, in my experience, some of these start-ups have great ideas, but if they are applying the technology to an industry with which they are not familiar; they may make some bad assumptions about your business operations. This is a rapidly evolving technology and its true, sometimes you will see some solutions in search of some problems, but other times you might be rewarded if you can help a vendor, especially a start-up better

© Springer Nature Switzerland AG 2021
G. R. Gray, *Blockchain Technology for Managers*,
https://doi.org/10.1007/978-3-030-85716-5_20

understand your needs. Let us look at a couple of bad assumptions that I have seen vendors make.

Bad Vendor Assumptions

The two biggest vendor assumptions regarding DLT-based solutions are:

1. Making assumptions about the computational capability of devices that may participate in a P2P network
2. Making assumptions about available bandwidth

These bad assumptions are usually seen in the start-up space. (Making my own assumption) I think that this is primarily due to start-ups being located in large cities or metropolitan areas such as New York or Silicon Valley. There are assumptions of ubiquitous broadband and everything runs as an app on your smartphone, (which as we have said earlier is really like having a supercomputer in your pocket). If it runs on a smartphone, it should run on everything, right? Now, if your solution only runs in the back-office, you *can* bet on lots of broadband and computational capability – just do not count on it if the back-office solution has to reach remote devices. Does your solution have a contingency for when remote devices have lost their connectivity?

An example of connectivity gone awry that you want to avoid was when a tech reporter from the Guardian (also, if possible, avoid customer SNAFUs with tech reporters – I do not see how that turns out well) was using a GiG car share vehicle. And it was great until the car was driven to a place where it was no longer receiving a signal… so it refused to move.[1] This situation was made worse with some poor customer service on the part of GiG and the fact that the reporter could receive a signal (thus facilitating the bad customer service with the GiG rep), although the vehicle could not. That the vehicle was always going to have connectivity was probably a bad vendor assumption.

When you get out of the city, or away from the interstate system (along which cell towers are located), things can still be a bit dicey from a bandwidth perspective. If the DLT transactions rely on connectivity to the Internet, or worse, rely on broadband connectivity to the Internet, that solution is not going to work in flyover country. I have actually seen a look of surprise on people making a presentation when they find out that outside of the Valley there are places where you cannot get a cell phone signal.

The other challenge is the computational capability assumptions of devices. Now, if the solution runs in the data center of the organizational, more than likely the servers (and the network) can handle all of the computational and bandwidth

---

[1] Driver stranded after connected rental car cannot call home, https://arstechnica.com/cars/2020/02/driver-stranded-after-connected-rental-car-cant-call-home/.

requirements. This is less likely for distributed devices. Smaller devices, even smaller than that pocket supercomputer, such as security cameras, locks, or sensors such as those that might be deployed to monitor the electrical grid, or even meters (electric, gas, water, steam), have limited computational capability. They often do not have the means to encode/decode hashes and have limited storage capability. Consider the current generation of the Raspberry Pi.[2] The current generation of these super small computers only cost $30 for the board. A little more for the case, power source, and displays, etc. These have roughly two to three times the computational power of the average electric meter. In a study conducted by the Electric Power Research Institute, to run an Ethereum-based energy trading simulation, the Pis had to run on their own private blockchain because even these little workhorses could not run the full blockchain.[3]

The computational consideration is an important consideration because a vendor either wants to (a) deploy a dApp on an existing legacy device, it needs to be able to run it – you will need to know what those computational requirements are, or (b) you will need to deploy devices that can run the dApp. And if you are only deploying a handful, the expense might not seem like much, but let us say you want to replace all of the electric meters in a territory, the cost will quickly scale to millions and millions of dollars. And that might be fine – if you can recoup that cost over the lifetime of the asset. Recall that this issue was something that we explored in the metering business case.

Bonus bad assumption:

- "It's blockchain so you should buy it."

- That is not a sales pitch. Now that we have moved a little further along the hype cycle that is not really a thing any longer – if it ever was. Remember, you have to come to the buyer with a story about how the solution provides a better, faster, cheaper way of doing something – or adding a new capability that did not exist before. What your solution is built upon may have very little to do with the core value proposition.

Bad Buyer Assumptions

You are not going to make any right? Because you read this book, right? But just in case let us review some of the potentially bad assumptions you might have made and have now avoided.

- It uses distributed ledger technology so it must be secure.

---

[2] https://www.raspberrypi.org/.

[3] Blockchain Market Simulation Using Ethereum: "Blockchain Village" https://www.epri.com/#/pages/product/000000003002015175/?lang=en-US.

Fig. 20.1 You are ready!

- As we learned, solutions based on DLT were some of the first to have security built-in "from the ground up" and not just lip service to the importance of security. But with the various hacks documented in this book you now know that the ecosystem of computer systems surrounding the core ledgers is just as vulnerable as legacy systems
- Cryptocurrency (digital money) spends just like real currency.
- Assuming that you can find a vendor that takes digital currency. But digital currency comes with its own problems – the most significant being the double-spending problem. Since the currency is digital and not a physical asset, many hoops are jumped through to prevent the double-spend. As a buyer you will want to know how the vendor's solution prevents this.
- "It's blockchain, so I should buy it."

- Now you have learned there are a lot of wrinkles to this technology, pros/cons of various architectures and consensus algorithms. You will still need to do you due diligence to ensure that a proposed solution meets your organization's needs. We have seen technology "silver bullets" come and go for decades. This will not be the last. But now you are armed with the knowledge you need to ask informed questions. Think back to our conversation on sustaining versus disruptive innovation. Does the solution have a better business case than a legacy solution – that draws on existing competencies? Or does it completely eliminate a process? Does it add a new capability? If the answer is "yes" to these, then you might have found a winner (Fig. 20.1).

Questions

1. Broadband communications is available everywhere your solution needs it (T/F)?
2. If a proposed solution uses DLT, it must be better than a solution that does not (T/F)?
3. If it runs on a Raspberry Pi, it will probably run a DLT-based solution (T/F)?
4. Disruptive innovation is competence destroying (T/F)?
5. Sustaining innovation is competence destroying (T/F)?

Chapter 21
Irony of Trust

Learning Objectives
- Review permissioned versus permissionless consensus algorithms and the implications for how the systems themselves are trustworthy

Trusted versus trustless in the world of DLT: But which is more trustworthy? What trustless means and the environment that is created may be a bit counterintuitive. As we have reviewed with the various consensus mechanisms that were explored earlier in this book, that all DLT consensus algorithms, since they are working with digital currency, must put in place mechanisms to avoid double-spending. The currency is digital, so there is no physical limit on double-spending as one would have with physical currency (Fig. 21.1).

Recall the conversation about Byzantine Generals and how they could not be trusted, hence, the steps that were taken to guarantee message fidelity in the face of that trustless environment. The participants were not trusted, but the system is designed to work even in the face of bad actors. Since the system is designed to work even with bad actors, anyone is allowed to participate – even if they do not want to be identified; there is no vetting of participants. (Although as we have seen with wallet providers, they may still enforce identification due to the banking laws of the jurisdictions in which they do business.) This "all comers" policy is known as a "permissionless" approach to participation. Anyone can join – bad intentions and all.

At the other extreme is the approach that is taken with PoA. In a PoA system, since it is assumed that there will also be bad actors, a different set of steps are taken. In a PoA, only known, verified entities are allowed to participate. This then, is a "permissioned" approach to DLT. Only those that have taken these authentication steps are allowed. The trade-off is not only with trust, but also performance. PoW mechanisms take more energy to solve the mathematical puzzle as a way to dissuade bad actors. The other mechanisms (PoA, PoS, DAG) use much less energy and also can clear blocks of transactions much faster than PoW (Fig. 21.2).

© Springer Nature Switzerland AG 2021
G. R. Gray, *Blockchain Technology for Managers*,
https://doi.org/10.1007/978-3-030-85716-5_21

Fig. 21.1 Proof-of-authority – we do not trust *you*

Fig. 21.2 Proof-of-work – We do not trust *anyone*

This is somewhat ironic if you think about the drivers for creating Bitcoin in the first place. It was due in part, to the financial crisis, which was brought on by known, verified actors – institutions such as banks and insurance companies. Considerable trust had been placed within these organizations, and yet when it mattered that trust failed. Now consider implementations of DLT such as Bitcoin and Ethereum, which are permissionless systems that allow anyone to participate, or systems such as zCash and Monero that use zero-knowledge proofs to keep participants completely anonymous. But such systems cannot be trusted, can they? They need to be regulated and have government oversight and we certainly would not want anyone to be party to a transaction without the government having their hand out now would we? But wait… didn't we have government oversight of the financial industry leading up to the huge crisis? Maybe *more* regulation was not what we needed, but a system that is incentivized to avoid risk or only have the entities that are party to a transaction be the ones at risk. A level of risk that they can decide upon for themselves via the terms encoded in a smart contract; which is not to say that regulators have no role in a DLT future. Although people are generally allowed to enter into bad contracts, there could be value in having a trusted party (if you trust regulators), review the terms of a smart contract to make sure that it legit, or at least to make sure that it does not have the sorts of flaws that were explored in Chap. 8 (prodigal, suicidal, and greedy).

It is interesting to note that a decade ago when cryptocurrency was just beginning to emerge there were plenty of naysayers in the financial sector. But as faith in fiat currencies has waned (as nations such as the United States begin piling on debt) causing interest in disinflationary vehicles such as Bitcoin and Ethereum, bankers have begun changing their tune… as people begin moving their money from banks to cryptocurrencies.[1] Suddenly banks will not be using cryptocurrencies merely as vehicles to reduce or eliminate transaction fees amongst their exchanges (fees for them, not for you), but as a means to retain financial holdings of their customers. That cat might be out of the bag in the sense that wallets like Coinbase make it extremely easy to buy, sell, and trade cryptocurrency, but there may be a class of traditional customer that might feel safer know that their holdings are tied to their traditional, legacy banking provider (Fig. 21.3).

The old saying is that when you are hearing stock advice from the waiter, you might be in a bubble (crash of 1929[2]). And clearly we have seen the effects of currency speculation – just take a look at Dogecoin after Elon Musk tweeted about it (but then it crashed after he appeared on Saturday Night Live[3]). Of course, it helps to put the "plunge" in context (see figure below). The "plunge" is the right-most dip shown in the graph. Will Dogecoin continue its drop? Was it just a coincidence that

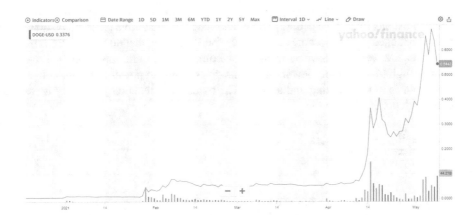

Fig. 21.3 Dogecoin financial performance history

---

[1] Bitcoin is coming to hundreds of U.S. banks this year, says crypto custody firm NYDIG, https://www.cnbc.com/2021/05/05/bitcoin-is-coming-to-hundreds-of-us-banks-says-crypto-firm-nydig-.html?utm_term=Autofeed&utm_medium=Social&utm_content=Main&utm_source=Twitter#Echobox=1620214102.

[2] Stock Market Crash of 1929, https://www.history.com/topics/great-depression/1929-stock-market-crash.

[3] Dogecoin plunges after Elon Musk mention on SNL, https://finance.yahoo.com/news/dogecoin-elon-musk-snl-cryptocurrency-bitcoin-ethereum-boe-andrew-bailey-091558489.html.

Elon said he bought it for his mom that caused the dip? Do you care about a crypto-currency that was created as a joke[4]? Since the writing of this book, Tesla has now said that it will no longer invest using Bitcoin because the carbon cost of the energy used is counter to the mission of Tesla – to provide cars that are a clean energy solution. And Elon claims to now be working with Doge to make it more efficient. There hardly seems to be a day that goes by without some announcement in the cryptocurrency space that does not make an impact in the market.

I do find it interesting to note that I have had conversations with technical people, but also creators such as musicians and graphic artists, where they do not discuss the "get rich quick" aspects of cryptocurrency; they discuss its long-term value and can go into detail about the differences between Bitcoin, Ethereum, and Ripple. When I have experiences like this, I cannot help but think cryptocurrency has arrived. People of all stripes can see the declining value of the dollar and are looking for ways (sometimes to get rich) to store wealth. Awareness has trickled down to the masses and the world will not be the same again. It was asked earlier in this book if DLT was a disruptive technology. Could something that underperforms incumbent financial security providers offer distinct characteristics that allow it to create its own market? I think the answer to the later question has become a clear and unequivocal, "yes". One thing that may make market watchers nervous as this new reality takes hold what happens to incumbents once the new technology entrants begin to perform on parity. The outcomes for incumbents usually do not turn out well. Just ask anyone at a company that was making eight inch disk drives when newly architected five inch drives came out. Oh wait, you cannot, because none of those companies exist any longer. To people that do not recognize disruptive technology, these market changes probably come as a big surprise. To people that do well, there is no guarantee that you are a savant, because sometimes the new entrants do not survive long enough to make their new markets that gives them the power to be disruptive. The proof is in the pudding and all that. The roadside of technology history is littered with better mousetraps.

Questions

1. Do you care that Dogecoin was created as a joke?
2. When a vendor proposes a DLT solution, not only must they address a pain point your organization has but must be able to describe the _____, _____, _____ of their solutions
3. A disruptive technology must do what to survive in the face of legacy competition?
4. What role can regulators play with smart contracts?
5. Is it a better mousetrap or is DLT just a blip on the technology radar?

---

[4] Dogecoin Started as a Joke, but It Has Serious Momentum Right Now, https://investorplace.com/2021/04/dogecoin-started-as-a-joke-but-it-has-serious-momentum-right-now/.

Part I Review

What a journey! But now you are ready to dig in deeper, yes? As you saw, DLT did not spring up "overnight" but has been created by numerous technologies to create a new capability and value proposition. Some of these technologies have been developed decades ago. But the coupling of these technologies into a set of core characteristics: security, privacy, immutability – are beginning to demonstrate the kind of changes that are associated with disruptive innovation. That does not necessarily mean that DLT will "win" in any given application – the "better, faster, cheaper" question still needs to be answered. But, now that you are armed with the understanding of the underpinning technologies of DLT, you will be better prepared to do your own evaluation of use cases based on this understanding.

© Springer Nature Switzerland AG 2021
G. R. Gray, *Blockchain Technology for Managers*,
https://doi.org/10.1007/978-3-030-85716-5

Part II Review

Part II covered the concept of immutability in greater detail. However, we have learned that immutability is not as unchangeable as one might have thought. This facet of DLT is one of its main value propositions – and a requirement for people to have faith that records are not changed and that money is not being double-spent, and in a word – reliable. If we are moving to where we do not need "trusted" entities like banks (that have had their own reliability challenges per the financial crisis), then the systems that are being used to replace them need to be reliable.

We also learned that while cryptocurrency applications are indeed built "from the ground up with security in mind", it does not mean that they are hack proof. The systems surrounding the core ledgers are subject to the same issues that all computers systems have suffered from: compromised passwords, privacy data, and phishing attacks.

In this section, we also took a deeper look into the Proof-of-Work process and some of the other alternative mechanisms and consensus schemes that are used for distributed ledgers. Each of these makes schemes make different design choices, and hence have different strengths and weaknesses that managers need to be aware of. These differences will become important as managers are asked to evaluate how applications, smart contracts, or mining services best fit solutions that they may build, buy, or invest in.

© Springer Nature Switzerland AG 2021
G. R. Gray, *Blockchain Technology for Managers*,
https://doi.org/10.1007/978-3-030-85716-5

Part III Review

Use cases and applications give you a chance to dig into how the core characteristics of DLT are applied in real-world settings. DLT is already starting to be a disruptive force in many market areas, but as we have seen from the section on these use cases, that does not necessarily mean it is a fit for all applications. DLT definitely seems to have an advantage anywhere there is a new area of application where no incumbent technology needs to be replaced (assuming that you do not think paper-based systems count as technology). It is often surprising where paper-based systems still exist. Are there paper-based systems where you work? This might be a great place to kick the tires on a DLT pilot. Often start-ups look to tackle BIG problems. And why would not you? Solving a BIG problem usually results in a fantastic reward.

But little problems have their own rewards that compound. By creating a small win and demonstrating the capability, the design team learns more about a business application and can demonstrate value early. Then the team can take this small win and build upon it. As we explored some of these use cases, you can see that many of them are parts of complex systems – some of which have been around for decades. While you might not be a fan of such systems, there is probably a lot of value locked into them. And when designing something that replaces them, you never want to be in a position where you are asking your employees or customers to step backwards in capability. Unless of course, your desire is to place yourself in a position of being asked to find a new mode of employment – well then, by all means.

And just for the record, if you can create a viable DLT-based voting system that restores confidence in the process – I for one would be a big fan of such an outcome.

G. R. Gray, *Blockchain Technology for Managers*, https://doi.org/10.1007/978-3-030-85716-5

Part IV Review

The demand catalyst for crypto is primarily a lack of faith in fiat currencies. That is the big one. The faith in it is based on the fact that it is distributed peer-to-peer network – it cannot be taken down or controlled by a government, for example, China with their currency manipulation. Nominally it is not "the best" crypto – "best" depends on the problem being solved. This was why other consensus algorithms were developed, usually to have faster "closing" times (how fast blocks are written) or block size. Many of the early coins were just forks of the Bitcoin code. The Ethereum network, for example, is better suited for distributed applications and hundreds have been developed for it. This is where I see the greatest disruption – the ability to have a common platform for the automation of business services that have resisted automation to date.

© Springer Nature Switzerland AG 2021
G. R. Gray, *Blockchain Technology for Managers*,
https://doi.org/10.1007/978-3-030-85716-5

Answers to Chapter Questions

Chapter 1

1. No one knows.
2. We cannot read Satoshi Nakamoto's mind, but probably the effects of the financial crisis played a part.
3. Again, we cannot read their mind, but government response to things viewed as within their purview is not usually magnanimous .
4. It depends – but companies have still been disrupted when they did.
5. It can be difficult, often disruptive technologies underperform relative to current market participants, but one clue is if they can create a new market based on their unique combination of characteristics.

Chapter 2

1. While there are more than four kinds of consensus mechanisms, the four main ones are PoW, PoS, PoA, and DAG.
2. False. The developer community may make changes that are both compatible and incompatible with prior versions which may fork the code.
3. False. The computer ecosystems around DLT are susceptible to the same vulnerabilities as any other type of computer system.
4. True.
5. Anonymous. Unless, of course, one uses a brokerage or exchange that requires you to identify yourself when using these types of cryptocurrencies.

© Springer Nature Switzerland AG 2021
G. R. Gray, *Blockchain Technology for Managers*,
https://doi.org/10.1007/978-3-030-85716-5

Chapter 3

1. The code base of Ethereum that was not altered to match its post-hack state.
2. Hard forks are not backwards compatible with previous versions of the code base. Soft forks are what you get when Neo is involved.
3. Ignore it .
4. No. It was a hack of an exchange.
5. The United States government.

Chapter 4

1. False
2. Proof-of-Capacity
3. Identified
4. True
5. Directed Acyclic Graph

Chapter 5

1. They exist electronically and their exchange is written to a blockchain, proving both uniqueness and ownership
2. True
3. Stable coin
4. The value of one of more fiat currencies such as the U.S. dollar, Euro, or yen.
5. The faith that people have in the government that backs them.

Chapter 6

1. The processing power of a crypto mining rig
2. Cost of power, cost of hardware, cost of HVAC
3. Mining reward payment schemes
4. It depends. Is the energy "clean" and do I care about the energy used to post pictures of food on Facebook and Instagram.
5. Is it the reuse of a brownfield or is it new development that may become a stranded asset?

Chapter 7

1. Security, transparency, immutability
2. Probably some combination of all three, recognizing that you cannot have it all
3. False. It means the likelihood of a block being changed rapidly approaches 0%
4. True
5. It was the case that this was true; only the systems in the ecosystem had been hacked, but now with the Ethereum ledger hack, we have seen an instance of the core blockchain contents being changed.

Chapter 8

1. An application that drives users to adopting the platform upon which it runs.
2. That the state of an application (data, step in a process, etc.) is kept track of.
3. Once it is stored in the ledger, that post is there forever.
4. An oracle facilitates data feeds onto a blockchain or the connection of multiple chains.
5. Prodigal, greedy, suicidal.

Chapter 9

1. There is really no wrong answer here. While some individuals companies may be changing or reducing their strategy, others see investment growing.
2. Current and long-term availability allowed to float with demand or tied to a fiat currency, a standalone asset or part of a software platform.
3. Implementation, sensitive information concerns, potential security threats, lack of regulatory clarity, challenges in forming a consortium, lack of in-house skills, uncertain ROI, lack of compelling application, unproven technology.
4. Mergers and acquisitions. Dominant players absorb smaller players.
5. The top three of this writing are Bitcoin (BTC), Ethereum (ETH), and Ripple (XRP).

Chapter 10

1. Where a single entity controls 51% (or more) of the nodes in a blockchain, thereby being able to determine which blocks of data are added to the chain – which could also mean prior blocks being overwritten (due to the longest chain wins rule).

2. No. The amount of time, vehicle, and payroll costs quickly destroys any business case where unattended operation is the expectation.
3. A group of computers under the control of another entity, usually for use in the creation of DDoS attacks.
4. Generally no. But they could run wallet applications.
5. Smartphone.

Chapter 11

1. There is no wrong question here. Auditors stake their reputation on their claims, but may lack the expertise that a professional hacker has. On the other hand, a hacker may not reveal all that they know.
2. A list of IP addresses that limit from where allowable network traffic can be received.
3. You would like to think so… but you cannot. Due diligence is required for all software components.
4. To track the settings and components in a system to be able to provide an audit trail of changes.
5. Physical security is still important. If someone has physical access to an electronic device, they could potentially compromise the communications to or from that device

Chapter 12

1. True
2. False
3. Clearing and settlement, payments, trade finance, identity, syndicated loans
4. Privacy concerns – you do not want that data falling into the wrong hands. Organizations that have that data can use it to determine what else you might buy, or what you are willing to pay for it – for good or ill
5. After the last Bitcoin is mined, you cannot mine any more. You can always plant more tulip bulbs.

Chapter 13

1. It can be used to trace where products and where their component parts come from. It can be used to determine if items are ethically sourced or trace defects or infection sources and corresponding distribution.

2. Radio frequency identification is a tag that has an antennae within it that facilitates being automatically scanned when placed into, or taken out of inventory, or any place along its journey from factory to use.
3. When goods have been diverted from their planned destination.
4. Probably not. But it looks promising that most could be eliminated.
5. Some reasons include many systems still use paper, with human-in-the-loop exchanges (that tend to be error-prone), and there are differences in how the associated platforms have implemented the exchange.

Chapter 14

1. A DER Management system
2. True
3. A form of DERMS that could be distributed to manage parts of the electrical grid
4. A person that may both buy and sell energy to any willing seller or buyer
5. Renewable Energy Certificate – used to track that generation of renewable energy

Chapter 15

1. Mostly corporate shenanigans
2. An immutable ledger whose contents cannot be modified – leaving an auditable trail of changes made or steps taken
3. Authorizing
4. Perhaps not – unless there was a clear return on investment for the new system, or concerns that somehow the existing system might be compromised by bad actors
5. People and equipment

Chapter 16

1. Broker
2. Hopefully not annoyed or surprised, but happy with the service that is provided with recourse with the original company if you are dissatisfied.
3. The people selling the shovels. The people providing the platform.
4. Broker or gig service. I would also accept "scam".
5. Of course. You are a forward thinker of excellent taste, that is not locked into old notions of what is, and what is not, suitable for pizza

Chapter 17

1. One is "dumb" and the other is "smart". Electromechanical meters need to be read by humans. Smart meters can be read remotely and automatically and may have additional features
2. 70%
3. 12–15 years
4. Mostly it comes down to bandwidth and data storage
5. No. The ledger is immutable. Other means would need to be found to correct for bad data

Chapter 18

1. True
2. Delicious. And also a small piece of data for tracking users on web sites
3. Garbage in-garbage-out. The ledger stores whatever is written to it – whether it is accurate or not
4. It requires that the government must receive the consent of the customer before they can access a customer's financial information
5. That datasets involving customers have to contain at least 15 customers, and no single customer's data can comprise more that 15% of the whole

Chapter 19

1. Innovation leads standards, but it often springs from standards that had been previously developed.
2. Everybody has one.
3. Internet Engineering Task Force.
4. Institute of Electrical and Electronics Engineers.
5. No. There are plenty of examples of successful products that were successful in spite of being conformant with a standard, for instance, Microsoft Office. But for many endeavors, the lack of standards compliance will eventually put a drag on performance if the product needs to be integrated with anything else

Chapter 20

1. Maybe. But better make sure before you sign on the dotted. Another option might be to determine if the solution will work if it is out of communication

2. Not necessarily. Focus on solutions that solve problems – not technologies
3. Probably – unless the storage requirements exceed the Pi's capacity. Make sure that you have planned for data growth.
4. Yes it is. Disruptive innovation fundamentally alters how work is performed.
5. False. Sustaining innovation builds on preexisting capabilities.

Chapter 21

1. I do not know about you, but it is kind of weird that a joke can be worth more than Ford Motor Co.
2. Better, faster, cheaper.
3. Carve out a market niche.
4. Reviewing the terms of contracts and making sure that they are not flawed.
5. It looks like DLT is here to stay – will it continue to disrupt financial and other companies? That remains to be seen.

Definitions

AMI	Advanced Metering Infrastructure: this refers to the more advanced type of smart meters that can automatically and remotely be read, but can also receive commands to initiate more advanced capabilities
AMR	Automated Meter Reading: this is the first generation of smart meters that could be read remotely.
Botnet	A "bot" is slang for a computer that is at the control of another party, either intentionally or unintentionally. A botnet is a group of these devices that are connected to the Internet that can be used to orchestrate cyberattacks.
Block	A block is a group of transactions. The size of the block determines how many transactions that it can hold.
Black Hat hacker	AKA "the bad guys".
DAO	Decentralized Autonomous Organization: it is the governance mechanism, via code, that runs the blockchain; it codifies the rules which are run by a software program.
dApp	Distributed application. These are based on smart contracts, with the transactions stored in a distributed ledger.
Distributed Denial of Service Attack (DDoS)	An attack that comes from many computers on the network (distributed), that attempts to overwhelm the target computer(s) with requests. The flood of requests prevents legitimate network traffic from reaching the source and may even cause the target computers to crash, thereby denying service to legitimate users.
DLT	Distributed Ledger Technology. There are various forms of DLT, some of which use a blockchain, but some use other means for maintaining a distributed ledger. Hence, DLT more accurately encompasses those distributed ledgers that use blockchain and other means.

© Springer Nature Switzerland AG 2021
G. R. Gray, *Blockchain Technology for Managers*,
https://doi.org/10.1007/978-3-030-85716-5

Double spending	Using the same money to purchase, or be party to, more than one transaction. If you spend $1 on item *a*, you cannot spend that same dollar on another item. This is one of the core transactional problems with digital currency that DLT attempts to solve for.
Fiat money	Is a currency that is not backed by a physical commodity such as gold or silver; has no intrinsic value other than that which has been assigned to it by the supporting government.
Fork	This is a reference to the software changes. Software that runs the various cryptocurrencies is kept in a code repository. Changes are managed via a change control system. When a change is made that is backwards compatible (it works with prior versions), it is considered a minor fork. When a breaking change is made (it does not work with a prior version of the software), this is a major fork. Most software follow a versioning numbering scheme of *major version.minor version.build* – For example, as of this writing, the version of Bitcoin Core is 0.21.1.
Hard Fork	This is a "breaking" change to the code base (the software) that runs a DLT. Each node in a distributed network needs to be on the same version of the software or else the node will not be able to correctly validate blocks, and hash transactions into valid blocks.
Mining Pool	It is not very feasible for a single person with a single (or even a few) desktop or graphics processing unit (GPU) to mine digital currencies. Thus, mining pools have sprung that allows individuals to join the pool, adding their computational power; the rewards earned by any individual are then shared across the members of the pool.
Nonce	A "number only used once". This is the number that is added to the hashed blocks in a blockchain. This is the number that the miners are attempting to solve for.
Permissioned/ Permissionless	This refers to the access to DLT. With a "permissionless" DLT, anyone can participate. In a "permissioned" DLT, only authorized and verified entities can participate. The permissionless form is most often associated with PoW-type DLT, while permissioned is most often associated with PoA.
Phishing	A cyberattack that involves communication that appears to come from a trusted source in an attempt to get the recipient to reveal their PII.
PII	Personal identifying information, for example, name, address, phone number, username/password
Smart contract	This is a contract that is automated. It is automatically executed once the terms of the agreement have been met. Smart contracts existed prior to the development of DLT.

Solidity	Ethereum built-in unique programming language
Token	For a given DLT, this is the currency that is used to represent value. For example, for Bitcoin, it is bitcoin, which uses the symbol BTC, for Ethereum, it is ether, with symbol ETH.
Wallet	A wallet, or more appropriately, "digital wallet" facilitates holding and transacting digital currencies.
White Hat Hacker	nominally a hacker that only uses their powers for good. (Wink, wink).

Index

© Springer Nature Switzerland AG 2021
G. R. Gray, *Blockchain Technology for Managers*,
https://doi.org/10.1007/978-3-030-85716-5

Printed in the United States
by Baker & Taylor Publisher Services